Fundamentals of Organic Chemistry

Fundamentals of Organic Chemistry

June Foster

STATES
ACADEMIC PRESS
www.statesacademicpress.com

Published by States Academic Press
109 South 5th Street,
Brooklyn, NY 11249, USA
www.statesacademicpress.com

Fundamentals of Organic Chemistry
June Foster

International Standard Book Number: 978-1-63989-229-7 (Hardback)

Cataloging-in-Publication Data

Fundamentals of organic chemistry / June Foster.
p. cm.
Includes bibliographical references and index.
ISBN 978-1-63989-229-7
1. Chemistry, Organic. 2. Chemistry. I. Foster, June.
QD251.3 .F86 2022
547--dc23

Contents

Preface

The branch of chemistry which studies the structure, properties, composition, reactions, and preparation of carbon-containing compounds is known as organic chemistry. These include hydrocarbons along with compounds with any number of other elements, including hydrogen, nitrogen, oxygen, halogens, phosphorus, silicon, and sulphur. Some common compounds studied under this branch are metalloids, transition metals, hydrocarbon compounds and organometallic compounds. The study of organic reactions includes the chemical synthesis of natural products, polymers, and drugs. It also involves the study of the individual organic molecules through laboratory methods and theoretical study. Organic chemistry finds its applications across various industries such as petroleum, biotechnology, fast moving consumer goods and pharmaceutical products. Organic chemistry is an upcoming field of science that has undergone rapid development over the past few decades. While understanding the long-term perspectives of the topics, the book makes an effort in highlighting their impact as a modern tool for the growth of the discipline. Those with an interest in the field of organic chemistry would find this book helpful.

A detailed account of the significant topics covered in this book is provided below:

Chapter 1- The branch of chemistry which deals with the study of structure, properties and reactions of organic compounds is known as organic chemistry. There are various subdisciplines of organic chemistry like stereochemistry, medicinal chemistry, computational chemistry, etc. This chapter serves as an introduction to organic chemistry.

Chapter 2- Organic compounds are the chemical compounds which contain carbon-hydrogen bonds. Organic compounds are classified into acyclic and cyclic compounds. Cyclic organic compounds are further divided into homocyclic and heterocyclic compounds. The topics elaborated in this chapter will help in gaining a better perspective about the classification of organic compounds.

Chapter 3- The chemical reactions involving organic compounds are organic reactions. There are many basic organic chemistry reactions like substitution reactions, addition reactions, elimination reactions, pericyclic reactions, rearrangement reactions and redox reactions. This chapter discusses in detail various types of reactions.

Chapter 4- Organic synthesis is concerned with the strategies involved in the intentional construction of organic compounds from easily available starting materials. Main areas of organic synthesis include total synthesis, semi-synthesis and methodology. Science and technology have undergone rapid developments in the past decade which has resulted in the discovery of significant tools and techniques in the field of organic synthesis, which have been extensively detailed in this chapter.

Chapter 5- Aromaticity, Covalent Bonding, Functional Groups and IUPAC Nomenclature are some of the basic concepts of organic chemistry. Aromaticity is the property of cycloalkenes where stabilization of the molecule is enhanced due to the ability of the electrons in the pi orbitals to delocalize. The topics elaborated in this chapter will help in gaining a better perspective about concepts of organic chemistry.

Chapter 6- An acid is hydrogen-containing molecule which has the ability to donate a proton to another molecule while base is any molecule which is capable of accepting a hydrogen ion from acid. Acid-base reaction is a chemical reaction in which exchange of hydrogen ions takes place between an acid and a base. This chapter delves into the diverse concepts and theories of acid-base chemistry.

I would like to make a special mention of my publisher who considered me worthy of this opportunity and also supported me throughout the process. I would also like to thank the editing team at the back-end who extended their help whenever required.

June Foster

1

Organic Chemistry: An Introduction

The branch of chemistry which deals with the study of structure, properties and reactions of organic compounds is known as organic chemistry. There are various subdisciplines of organic chemistry like stereochemistry, medicinal chemistry, computational chemistry, etc. This chapter serves as an introduction to organic chemistry.

Organic chemistry is the branch of chemistry that involves the scientific study of organic compounds (compounds that contain covalently bonded carbon atoms). This branch of chemistry primarily deals with the structure and chemical composition of organic compounds, the physical and chemical properties of organic compounds, and the chemical reactions undergone by these compounds. Advancements in the field of organic chemistry have made numerous contributions to human society, such as the synthesis of several drugs, polymers, and other natural products. Synthetic organic chemistry is an important application of organic chemistry which deals with the design and construction of organic compounds for practical purposes.

The term 'organic' was coined because the field of organic chemistry was initially finite to compounds that were produced by the living organisms. This was attributed to some 'vital force' that was present in organic substances as they had something that inanimate substances lacked. The above theory was debunked when Urey Miller synthesized urea from inorganic substances but the classification is still in use.

Organic chemistry is a vast science due to one key property exhibited by the element carbon, which is called carbon catenation. Carbon has the exceptional ability to form very stable bonds with other carbon atoms, giving it the ability to form stable molecules with relatively complex structures. Catenation is the ability of an element to form bonds with an atom of the same kind. Hence, the vastness of organic chemistry can be attributed to this property of carbon.

The importance of organic chemistry in the present age is as immense as it had been since its inception. It plays an important role in our everyday life because food, medicines, paper, clothes, soap, perfumes, etc are indispensable to us for proper living. The study of organic chemistry is important for chemists and pharmacists in synthesizing medicines for the alleviation of human suffering. The reactions in organic chemistry occur between organic compounds.

Cleavage of Bonds

The reactions in organic chemistry occur by the breaking and making of bonds. Bonds can cleave in either of two ways:

- Homolytic Cleavage.
- Heterolytic Cleavage.

Homolytic Cleavage

If the covalent bonds between two elements break in such a way, that each of the elements gets their own electrons, it is called homolytic cleavage. That is, each element gets an electron. Homolytic cleavage results in the formation of free radicals.

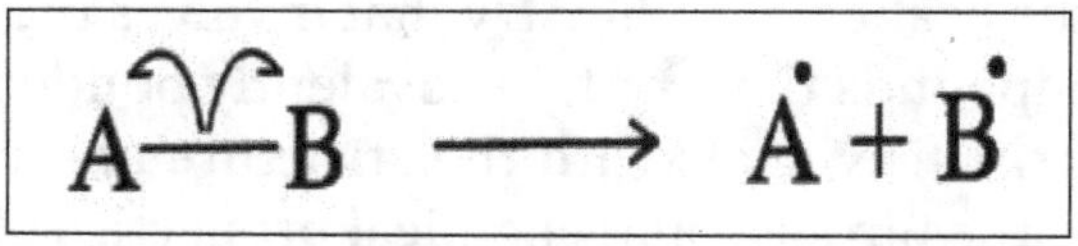

Homolytic Cleavage of Covalent Bond.

We have used an arrow to show the movement of electrons. Here, in this case, the arrow used is called the fish-hook arrow, as it signifies that there is a movement of only one electron.

Heterolytic Cleavage

If the covalent bonds between two elements break heterolytically, i.e., unequally, it results in the formation of charged species. This type of bond breaking, where the electrons are unevenly distributed is called heterolytic cleavage.

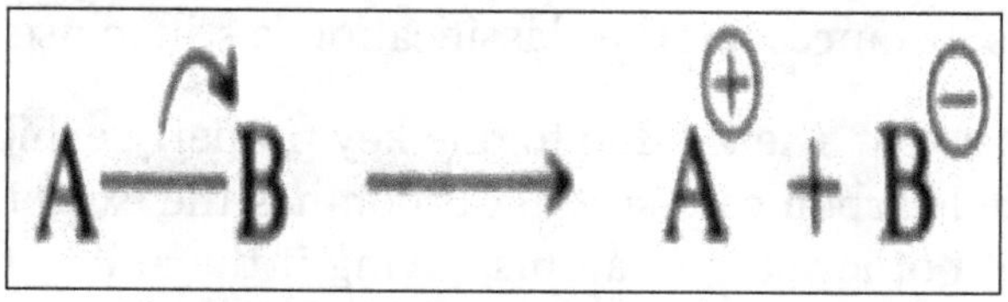

Heterolytic Cleavage of Covalent Bond.

In the above figure, we have used arrows to signify the movement of electrons, a regular arrow signifies that two electrons are being moved.

Reaction Intermediates in Organic Chemistry

Intermediates can be understood as the first product of a consecutive reaction. For example, in a chemical reaction, if A→B and B→C, then, B can be said to be the intermediate for the reaction A→C. The reactions in organic chemistry occur via the formation of these intermediates.

Carbenes

Carbenes (H2C) are neutral and reactive species that have six electrons in the outer shell of carbon making them electron deficient. Since carbenes are species having two odd electrons, we can classify carbenes based on their spin states.

Singlet Carbene

The electrons are present in different orbitals with opposite spin. The electrons are paired in sp^2 hybridized orbitals and behave as paired electrons.

Spin state= (2S + 1), S for singlet carbene is zero as the electrons are antiparallel. Therefore, Spin state = (2 × 0 + 1) = 1.

Triplet Carbene

Both the electrons are present in different orbitals and they possess the same spin. Spin state = (2S + 1), S for triplet carbene is 1 as both the electrons have the same spin. Therefore,

Spin state= (2 × 1 + 1) = 3

Hybridization of Singlet and Triplet Carbene

Singlet carbene Hybridisation: They are sp^2 hybridized with a bent shape. They have a bond angle of 103° and bond length of 112 pm.

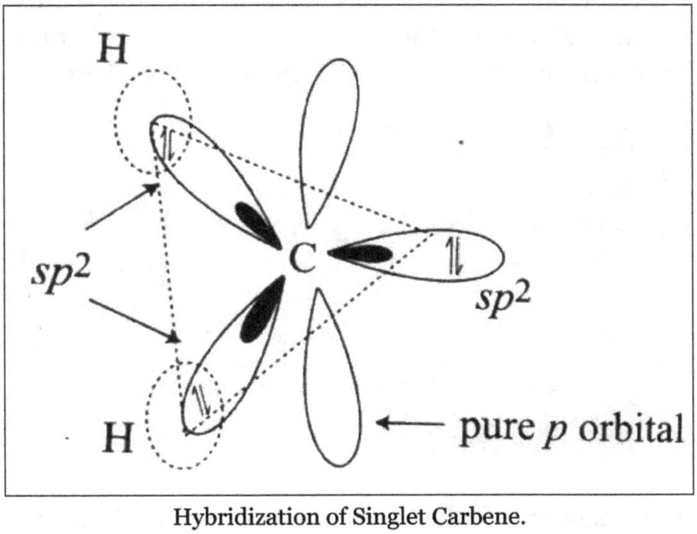

Hybridization of Singlet Carbene.

Triplet Carbene Hybridization: They possess sp hybrid orbital with a linear shape. They have a bond angle and bond length of 180° and 103 pm respectively.

Triplet Carbene is more Stable than Singlet Carbene

Triplet carbene has lower energy then singlet carbene because in singlet carbene there are more inter-electronic repulsions as both the electrons exist in the same orbital whereas in triplet carbene the two electrons exist in different orbitals making it possess less energy.

Free Radicals

Free radicals in organic chemistry are formed by the homolytic cleavage of carbon bond. The shape of the species formed is planar and the carbon is sp^3 hybridized with an odd electron being placed in the p-orbital. If the free radical is relatively stable then they may possess a planar structure.

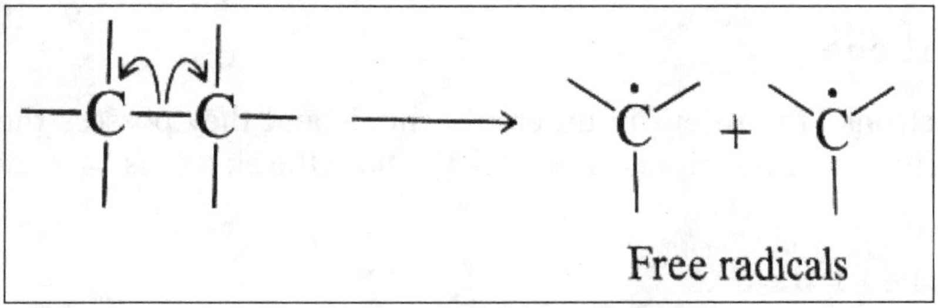

Planar Structure of Free Radicals.

Carbanions

They are generated by heterolytically cleaving a group attached to carbon without removing the bonded electrons. This makes the carbon have a pair of electrons, thereby, imparting a negative charge on the carbon. CH_3^- is isoelectronic with NH_3 and it is sp^3 hybridized and the shape is pyramidal owing to the presence of a lone pair of electrons.

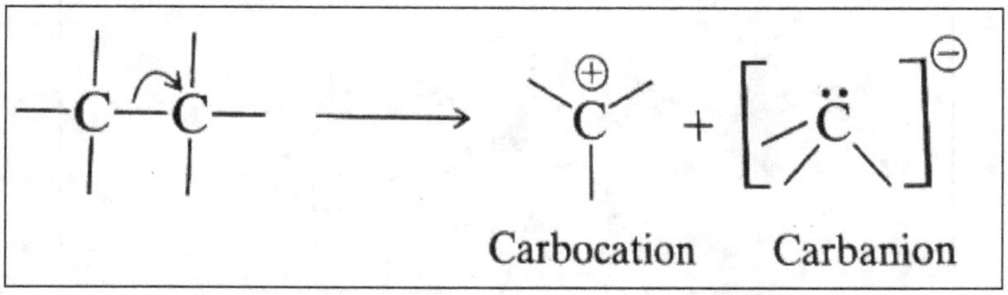

Formation of Carbanions and Carbocations.

Carbocations

Carbocations have a sextet of electrons on the carbon-containing the positive charge and hence termed 'cation'. It is sp2 hybridized and has an empty p-orbital. The shape is planar. It is generally formed by heterolytic cleavage of a carbon-heteroatom bond.

Transition State in Organic Reactions

The intermediates that could be formed in an organic reaction, now let us look into transition states and the difference between an intermediate and a transition state.

The intermediates in organic chemistry are formed in a multi-step reaction but some reactions can occur in a single step without having to form an intermediate. These reactions will occur by going through a transition state. This can be clear by looking at the energy profile diagram for a reaction, R→P.

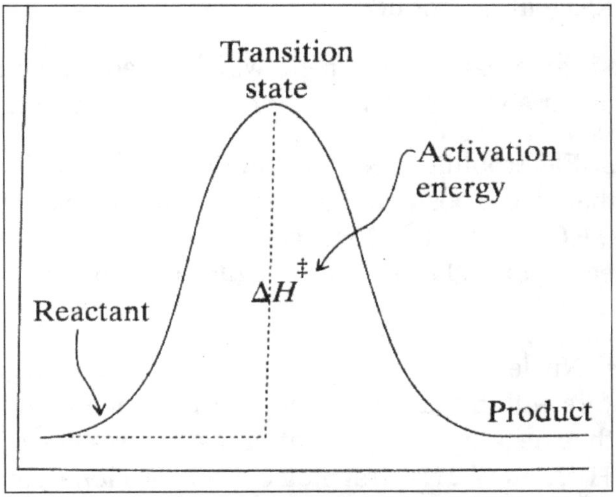

Energy diagram of single-step reactions.

The transition state, corresponds to the highest energy in the reaction, after which it can give either the products or in the case of a reversible reaction, the reactants. Consider a reaction, A→D with the following steps, A→B, B→C, and C→D. The energy profile for this reaction is given below.

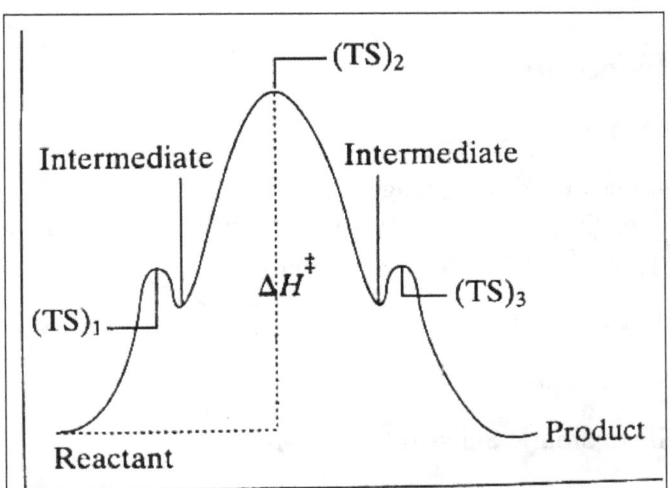

Energy profile of multiple-step reactions.

We can see that B and C are the products of a reaction and hence they are termed as intermediates. The highest energy of a particular reaction should be the transition state.

Reagents in Organic Chemistry

Reagents are the chemicals that we add to bring about a specific change to an organic molecule. Any general reaction in organic chemistry can be written as:

Substrate + Reagent → Product

Where the substrate is an organic molecule to which we add the reagent. Based on the ability to either donate or abstract electrons, the reagents can be classified as:

- Electrophiles: Electrophiles are electron deficient organic reagents. It can be generalized that all the positive charge containing species are electrophiles. For example, H+, NO_2^+, CH_3^+, Cl^+. Neutral molecules that are electron deficient can also act as electrophiles. Lewis acids like AlCl3 and BF3 are examples of neutral electrophiles.
- Nucleophiles: Nucleophiles are electron rich organic reagents. They seek for bonding centers with other nuclei and hence the name, nucleophile. It can be generalized that negative charge containing species are nucleophiles. For example, H^-, CH_3^-, and Cl^-. Neutral molecules with a lone-pair of electrons on the heteroatom can act as a nucleophile. For example, H_2O, NH_3, CH_3OH.

Types of Reactions in Organic Chemistry

Organic reactions are reactions that occur between organic compounds. The reactions in organic chemistry are broadly classified into six categories.

Substitution Reactions

R-X + Y → R-Y +X

Where R-X is the substrate, Y is the reagent (which can be electrophilic or nucleophilic) and X is called the leaving group. The term substitution means one group is replacing the other group. Types of substitution reaction are:

- Nucleophilic Substitution (S_N1, S_N2, S_Ni).
- Electrophilic Substitution (S_E).
- Nucleophilic Aromatic Substitution (S_NAr).
- Addition Reactions.

Addition reactions can be further classified into:

- Electrophilic Addition.
- Nucleophilic Addition.
- Elimination reactions.

Elimination Reactions

These reactions can be said to be the reverse of an addition reaction, wherein a simple molecule (HX, H_2O) is removed from the substrate i.e. a molecule is said to be eliminated from the substrate. Elimination reaction can be classified further into E1, E2, E1CB.

- Oxidation and reduction reactions.
- Pericyclic reactions.
- Molecular rearrangements.

Field Effect in Organic Chemistry

Inductive Effect

It is an electron delocalization effect via σ bonds that arises due to the difference in electro-negativities. For example, in a σ bonded organic compound like C-C-C-Cl, the carbon attached to the chlorine atom can be referred to as the α-carbon, and the one adjacent that carbon as the ß-carbon and so on.

Now, since Chlorine is more electronegative than carbon, it withdraws the electrons that are present via the σ bond toward itself, thereby making Cα fractionally positive. Since it is devoid of electrons, Cα now being slightly electropositive than Cß pulls the sigma bonded electrons of Cα-Cß bond toward itself and in this process, it makes Cß slightly electropositive.

The electron withdrawing effect of the Chlorine atom is being transmitted through the carbon chain via the σ bonds. This transmission of charges decreases rapidly with the number of intervening σ bonds. We can practically ignore this effect beyond Cß.

$$\text{-I-effect: } -\overset{|}{\underset{|}{C}}-\overset{|}{\underset{|}{C}}-\overset{|}{\underset{|}{C}}\rightarrow Cl \longrightarrow -\overset{|\,\delta\delta\delta^+}{\underset{|\,\gamma}{C}}-\overset{|\,\delta\delta^+}{\underset{|\,\beta}{C}}-\overset{|\,\delta^+}{\underset{|\,\alpha}{C}}-\overset{\delta^-}{Cl}$$

The arrow is pointed toward the more electronegative atom. If a group withdraws the electron from carbon, it makes carbon slightly electropositive. Such groups are called -I groups and the effect as -I effect. For example, -Cl, -Br, -CN and $-NO_2$ are -I groups. Groups that release electrons towards carbon are termed as +I groups and the effect is termed as +I inductive effect in organic chemistry. For example, alkyl groups like $-CH_3$ are +I groups.

$$\text{+I-effect: } -\overset{|}{\underset{|}{C}}-\overset{|}{\underset{|}{C}}\leftarrow CH_3 \longrightarrow \overset{|\,\delta\delta^+}{\underset{|}{C}}-\overset{|\,\delta^+}{\underset{|}{C}}-\overset{\delta^+}{C}H_3$$

Electromeric Effect

It is the temporary delocalization of π-electrons in a compound containing multiple covalent bonds. It is important to note that it is only a temporary effect, that is, it occurs only when a reagent is added. Electromeric Effect in organic chemistry can be classified into two types:

- Positive Electromeric Effect.
- Negative Electromeric Effect.

Positive Electromeric Effect

When the π-electrons are given to the attacking reagent. For example, the reactions alkenes and alkynes mostly occur via +E, this reaction is also called electrophilic addition.

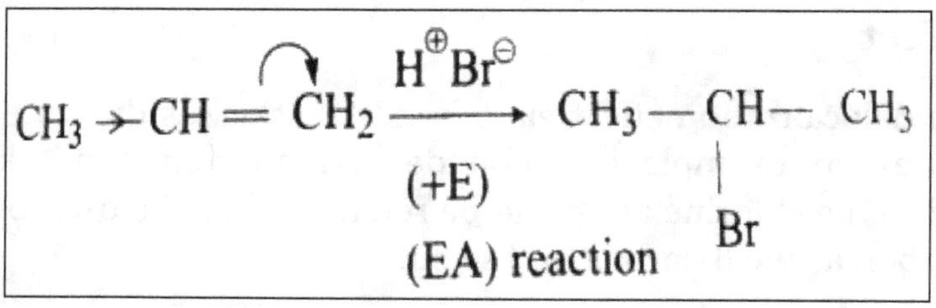

The proton adds at C-1 as the π-electrons were given to the attacking reagent (H^+). This results in the formation of a carbocation.

Negative Electromeric Effect

When the π-electrons are shifted to a more electronegative atom (O, N, S) joined via multiple bonds. For example, the reactions of aldehydes and ketones occur predominantly by -E effect. It is also called nucleophilic addition.

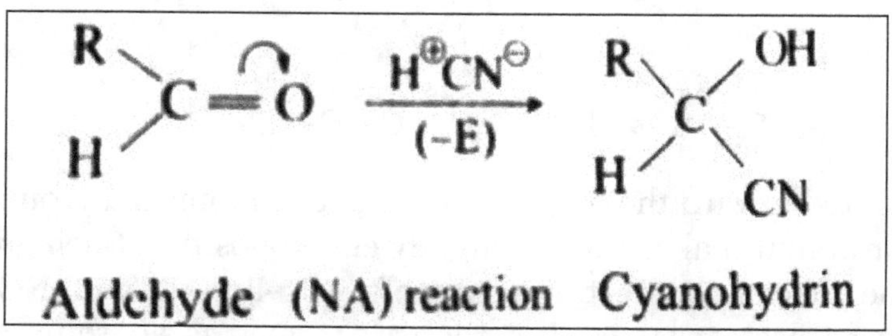

The CN- ion adds to the C atom of the carboxy group opposite to the movement of the π-electron cloud.

Mesomeric Effect

Molecules possessing sigma-bonds and pi-bonds alternatively exhibit the mesomeric effect. The effect is exhibited due to the permanent delocalization of π-bonds. This increases the number of resonating structures which makes the molecules of organic chemistry more stable. Such kind of a system where there are alternative sigma and pie bonds are called conjugated.

Types of mesomeric effect:

Positive Mesomeric Effect: This effect is exhibited when the direction of the delocalization of electrons is away from the position where the group is attached. Normally groups having a lone pair of electrons attached to a conjugated system push electrons into the conjugated system, that is, away from them. Groups in organic chemistry showing positive mesomeric effect (+M) effect are -OH, -OR, $-NH_2$, -SH, -X, etc.

- Negative Mesomeric Effect: This effect is exhibited when the direction of the delocalization of electrons is towards the position where the group is attached. These are generally electron withdrawing groups of organic chemistry.
- Resonance Effect: For certain molecules like carbonate ion (CO_3^{2-}), one single Lewis structure would not be enough to explain all of the properties. In that case, the molecule is said to have more than one structure.

Each of those structures can explain some of the properties but not all of the properties. The actual structure of the molecule is a hybrid of all the possible structures (canonical forms). This phenomenon is called resonance in organic chemistry. If resonance occurs, each bond would be both, a single bond and a double bond at the same time i.e. the bond order would lie between one and two. Resonating structures should fulfill the following criteria:

- All atoms should have the same positions in all the structures.
- There should have the same number of paired and unpaired electrons.
- The structures should have almost the same energies.

Canonical forms do not have any existence in reality.

Resonance Energy

The energy difference between the most stable canonical form and the resonance hybrid is known as Resonance Energy. More the resonance energy more is the stability. Rules for finding out the most stable canonical form:

- The canonical form with no charges is the most stable.
- The canonical form with more number of covalent bonds is more stable.
- The canonical form where unlike charges are in close proximity are more stable.
- If there are to be charged, the negative charge should be on an electronegative atom. Then this canonical form is said to be more stable.

Role of Steric Hindrance

The structure and reactivity of many compounds in organic chemistry are greatly dictated by the presence of bulky groups or constituents in the molecule. This is called steric hindrance. It arises because of inter-electronic repulsions due to spatial crowding amongst bulky groups. Using steric factors, we can conclude that trans-2-butene is more stable than cis-2-butene.

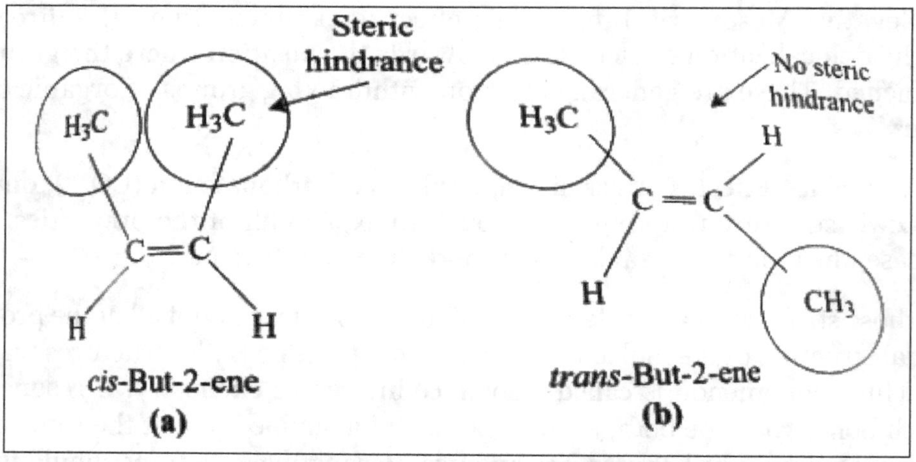

Steric Hindrance in Organic Chemistry.

Stability of Intermediates

Carbocations

The stability of carbocations can be explained by the Inductive Effect. CH_3^{+} $^{+}CH_2(CH_3)$ $^{+}CH(CH_3)_2$ $^{+}C(CH_3)_3$ i.e. Methyl Primary (1°) Secondary (2°) Tertiary(3°). We know that alkyl groups of organic chemistry are +I groups, that is, they release electrons through the sigma bonds.

Since the carbon is deficient of electrons, we can say that as the number of methyl group increases, the stability of the carbocation increases. Since the electropositive carbon is satiated by the electrons given by the methyl groups via +I effect.

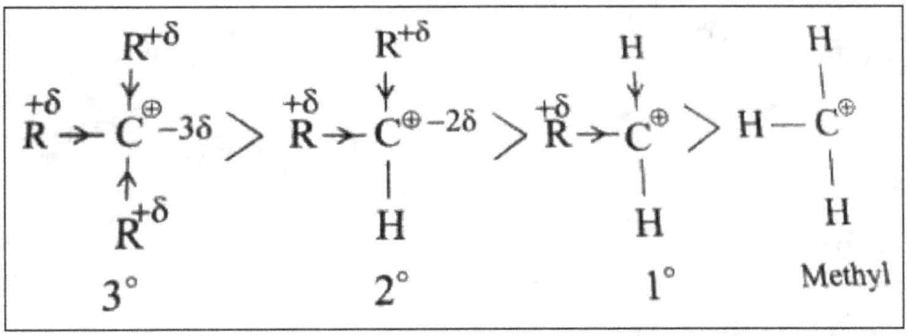

Therefore, the stability order of carbocations in organic chemistry is of the order 3° > 2° > 1° > Methyl carbocation.

Hyperconjugation

When a C-H σ-bond is in conjugation with a carbocation, this effect is observed. A carbocation has a vacant p-orbital. The bonded σ-electron pair of the C-H bond is displaced toward the vacant p-atomic orbital. This increases the electron density in the empty p-AO.

It is, therefore, a resonance effect where a C-H bond breaks and the σ-electron pair is delocalized to the vacant p-AO of the carbonation. Since the bond between C and H is broken, it is also called as 'no bond resonance'. It is also referred to as the Baker-Nathan effect.

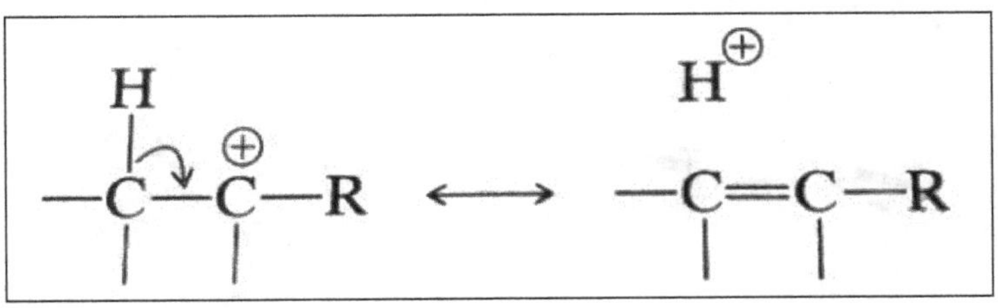

More the number of α-hydrogens in a carbocation, that many hyper-conjugated structures would be possible. More the number of structures more is the stability of the species.

Stability of Carbanions

The stability of carbanions can be explained using the inductive effect. $CH_3 - CH_2(CH_3) - CH(CH_3)_2 - C(CH_3)_3$ i.e. Methyl Primary (1°) Secondary (2°) Tertiary (3°). Since alkyl groups are electron releasing by nature through induction, we can say that more the number of methyl groups attached to the carbon having a negative charge, less would be its stability.

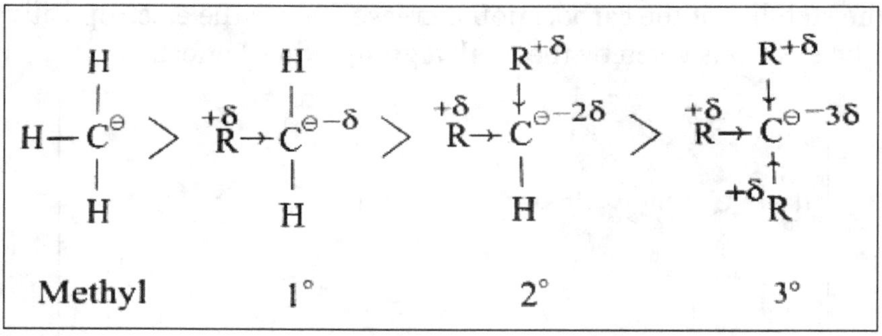

This is because, the carbon already has a negative charge, to which the methyl groups push electrons via induction. This results in inter-electronic repulsions and destabilizes the species. Therefore the stability order for carbanions is as follows: 3°< 2°< 1°< Methyl carbanion.

Stability of Free Radicals

The stability of free radicals in organic chemistry follows the same trend as that of carbocations. CH_3–$CH_2(CH_3)$–$CH(CH_3)_2$–$C(CH_3)_3$ i.e. Methyl Primary (1°) — Secondary (2°) — Tertiary(3°). Therefore, the stability order of free radicals is of order: 3° > 2° > 1° > Methyl carbocation.

This can be explained with the help of Hyper-conjugation that we saw, but here there would be an overlap between the σ-bond of C-H and the odd electron in the p-orbital of carbon.

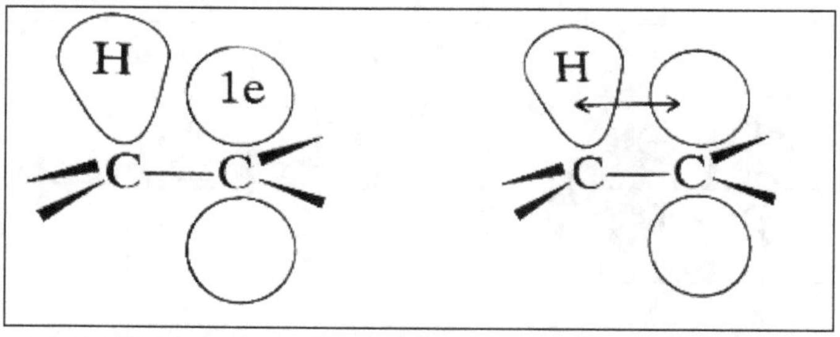

If there is a possibility of resonance, then that would make it more stable. This is because resonance affects the entire structure.

Branches of Organic Chemistry

Stereochemistry

Stereochemistry is the scientific concept that describes the relationship between chemical compounds which have the same molecular formula and identical functional groups but with different three dimensional arrangements of atoms. These molecules are known as stereoisomers and can have very different properties from one another, particularly when interacting with biological systems.

Enantiomers—Non Superimposible Mirror Images

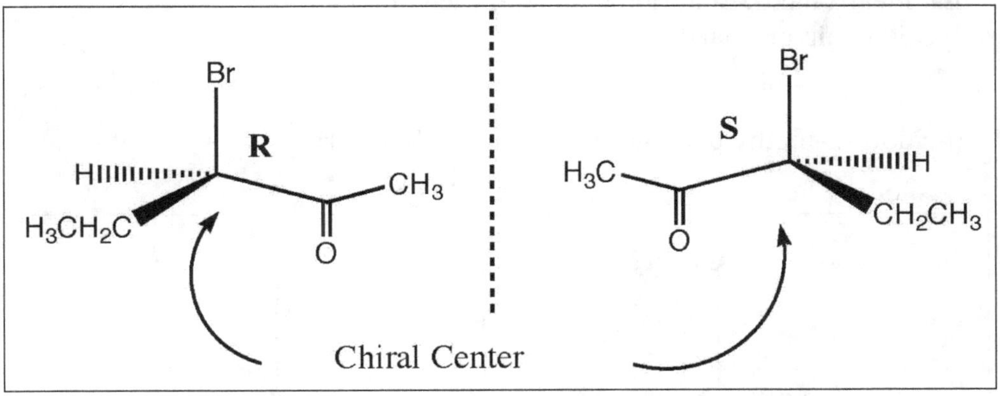

Enantiomers.

As shown the R and S enantiomers are together as an equal amount in a racemic mixture. They must be separated in order to study their stereochemical properties.

Configurational Isomers–Chirality

Another type of isomerization occurs when a carbon atom is bound to four different substituents. This is called configurational isomerism. Configurational isomers have as their only difference the way they are oriented in space, their three-dimensional arrangement. Although configurational isomers can be difficult to visualize and understand, they are extremely important especially in biological chemistry. For example, the pain reliever Ibuprofen exists as configurational isomers but only one isomer is effective (the S isomer) in treating pain. Also, the drug L-DOPA (Ldopamine) used for treatment of Parkinson's disease is effective only as the L or R isomer in the treatment.

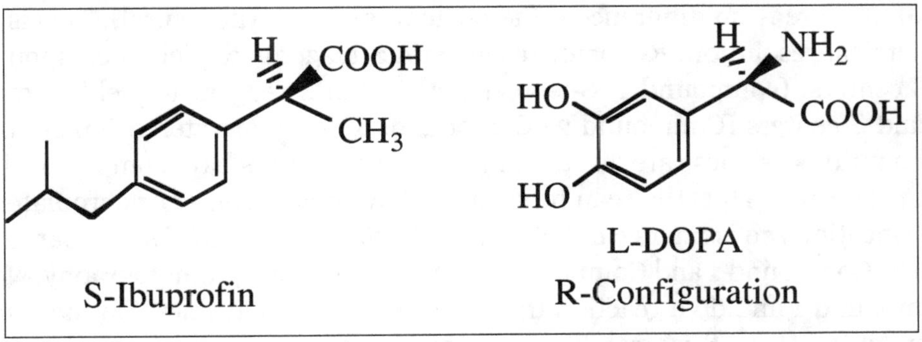

In 3-chloro-2-methylpentane, four different substituents are bound to carbon 3; an ethyl group, an isopropyl group, a chlorine atom and a hydrogen atom. When the compound is placed in front of a mirror, it presents a mirror image that is not superimposible on the original structure. This means that there is no way to orient the two structures such that they can be identical. The non superimposible mirror images are called enantiomers, which are configurational isomers. The enantiomers are also called chiral which means they lack symmetry. The word chiral comes from a Greek word chiros that means hand. Enantiomers are to each other as your hands are to you, non-superimposible mirror images.

3-chloro-2-methylpentane: 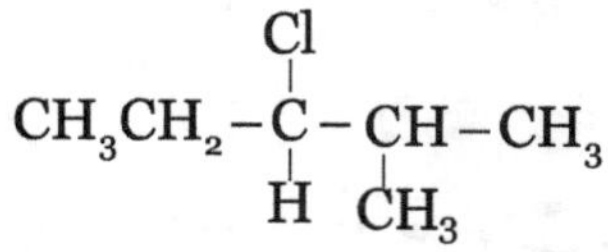

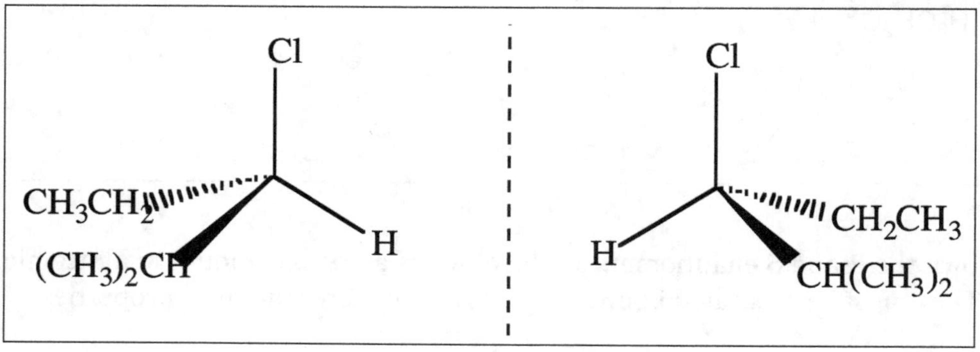

Enantiomers are identical in most properties such as melting point, boiling point, but they are different in the way they react with other chiral molecules and in the way they interact with polarized light. Their interaction with polarized light is called optical activity.

Optical Activity

Pure Enantiomers (The R and S are Separated from Each Other)

Will rotate plane polarized light (optically active). There experiments involve the use of Optical activity studies. Each enantiomer will react differently with another chiral reagent, but there is no difference if the reagent is achiral (not chiral). This is the basis of chemical resolution. R-isomer reacts with S-reagents to yield a compound with R and S centers. (compound 1) S-isomer reacts with S-reagent to yield a compound with S and S centers (Compound 2) Compounds 1 and 2 are stereoisomers but they are not enantiomers, these are diastereomers. Diastereomers have completely different physical properties. Thus they can be separated by crystallization or chromatography. After purification the isomers can be released back in their pure form separated from each other. Compound 1 and Compound 2 are separated by chromatography. Now take pure compound 1 and do a reaction that regenerates the R-isomer. Do the same with compound 2 to get the S-isomer. This is resolution of an enantiomeric mixture. Many natural compounds such as enzymes are not symmetrical. Thus individual enantiomers will react with enzymes to give different results.

A pure chiral compound, not a mixture of enantiomers, will interact with plane polarized light and cause the plane to rotate to the right (dextrorotarory) or to the left (levorotatory). An equal mixture of enantiomers will not show optical activity because half of the molecules would rotate light to the left while the other half would rotate the light to the right with a net rotation of zero. Thus enantiomers must be separated in order to observe optical activity. Optical rotation is an inherent property of an optically active compounds and is used as a physical constant for characterization of the compound. Optical rotation depends on the arrangement of atoms or groups around the chiral center—the configuration. Optical activity is measured automatically with an instrument called a polarimeter.

Naming Configurational Isomers

Configurational isomers contain carbon atoms with four different substituents. The carbon atoms are called stereogenic centers or chiral centers. A naming system has been devised so that we can distinguish one enantiomer from the other based on the orientation of those substituents.

The system is called the Cahn-lngold-Prelog method, and it follows several rules. First the substituents on the stereogenic carbon are assigned a priority based on atomic number. Low priority is given to low atomic number. If identical atoms are attached to the stereogenic carbon, then priorities are determined based on the atomic number of the next atom attached. Thus, in 3-chloro-2-methyl pentane, the lowest priority goes to hydrogen and the highest priority goes to chlorine (labeled 1). The carbon atoms of ethyl and isopropyl are identical so we go out one atom. Now the ethyl has one carbon (methyl) attached to the CH2 while the isopropyl has two carbons (methyls) attached to the CH. Thus the isopropyl group gets the higher priority (labeled 2) and the ethyl group gets priority 3. After the priorities are assigned, the molecule must be oriented with the lowest priority group pointing away from the observer.

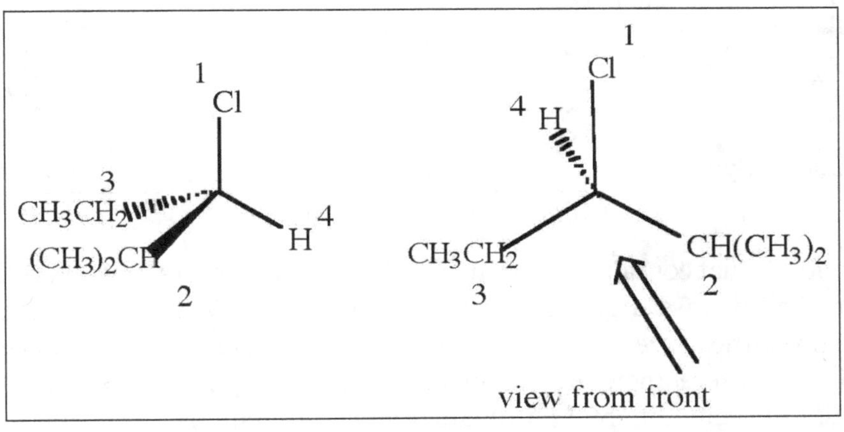

After getting the correct view, draw a circle from priority 1 to 2 to 3. If this circle makes a right hand turn then the configuration is called R. The R comes from the Latin word rectus, meaning right. If we draw a circle with a left turn then the configuration is S, meaning left which in Latin is sinister. A stereogenic carbon atom thus has two different designations, R or S, depending on the orientation of the substituents. In our molecule the configuration is R, and the complete name of this enantiomer is R-3-chloro-2-methylpentane.

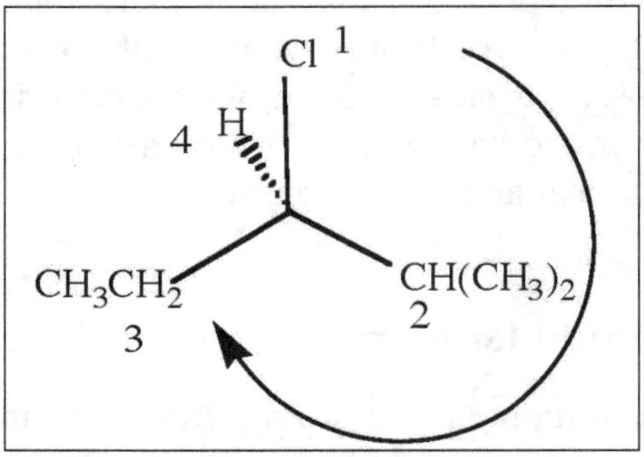

Right hand turn.

More than One Chiral (Stereogenic) Center

When molecules have more than one stereogenic center, structures such as dotted-line wedge and sawhorse may still be used, but another useful type of structure is the Fischer projection. In the Fischer projection all of the vertical lines go back into the paper and the horizontal bonds come out toward you. You have to remember this method because the structure looks flat. Also the only kind of change you can do to the structure is a rotation of 180°. All of these methods for writing stereo-chemical structures are important. The Fischer projection is used mostly in the study of carbohydrates.

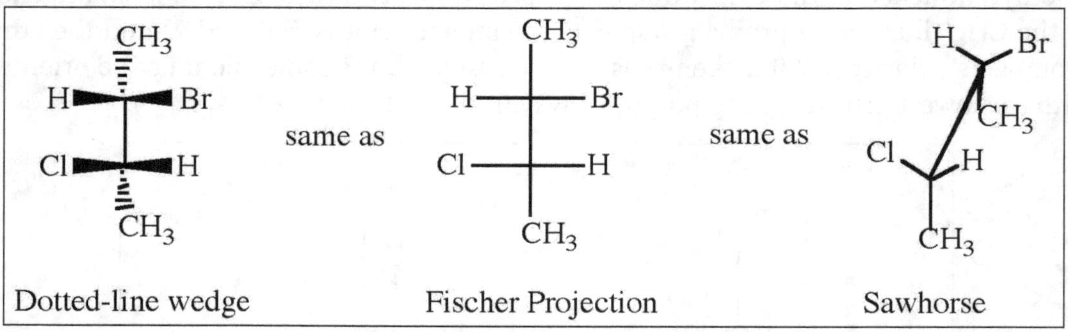

(2S, 3S)- 2-bromo-3-chlorobutane.

Many compounds that contain more than one chiral center are stereoisomers that are not enantiomers. In this case they are called diastereoisomers. While enantiomers have all of the same properties except for their interaction with other chiral substances and with polarized light, diastereomers are completely different in physical properties (different solubility, different mp, different chromatography, different spectra). The compounds 1 and 2 below are enantiomers as they are non-superimposible mirror images, but compound 3 is not an enantiomer of 1 or 2. Thus 1 and 3, and 2 and 3 are diastereomers.

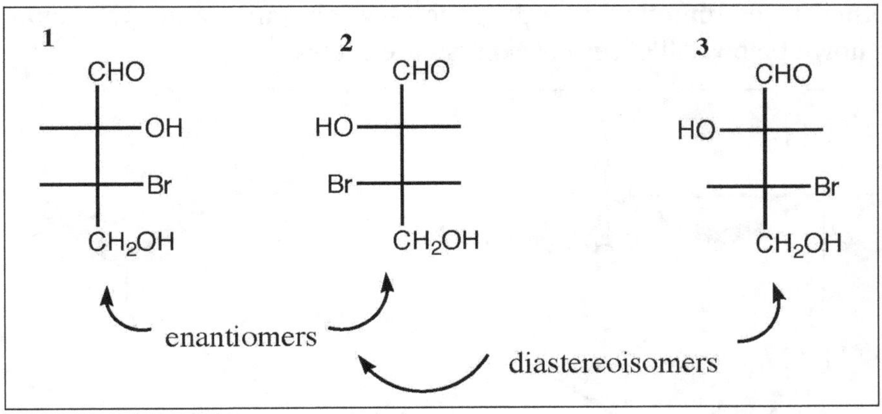

Chirality in Cyclic Systems

Stereogenic centers may also exist in cyclic systems. The cyclic structure must be tested for a plane of symmetry, and if one is present the molecule is not chiral. It is an achiral molecule designated as meso. The example below with cis and trans 1,3- dimethylcyclohexane shows that the cis isomer contains a plane of symmetry and is achiral. The trans isomer is chiral as it contains no symmetry planes.

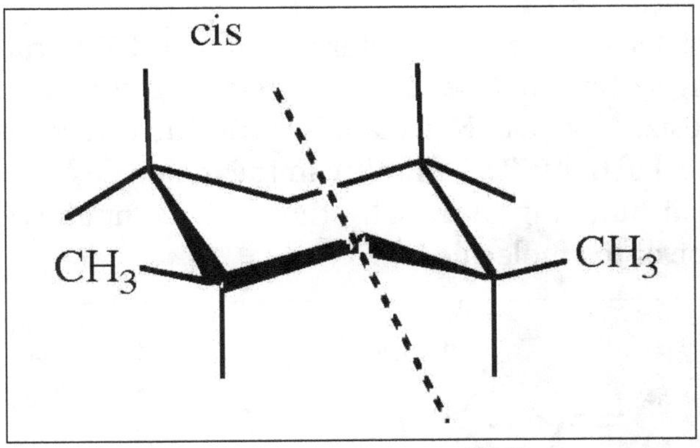

Plane of symmetry not chiral.

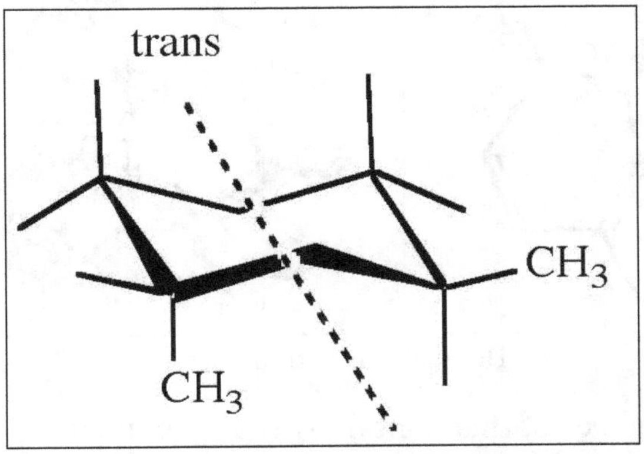

No plane of symmetry two chiral atoms.

Several other trans-dimethyl-cyclopropane, -cyclobutane, and-cyclopentane chiral rings are shown below. All of the cis isomers are achiral.

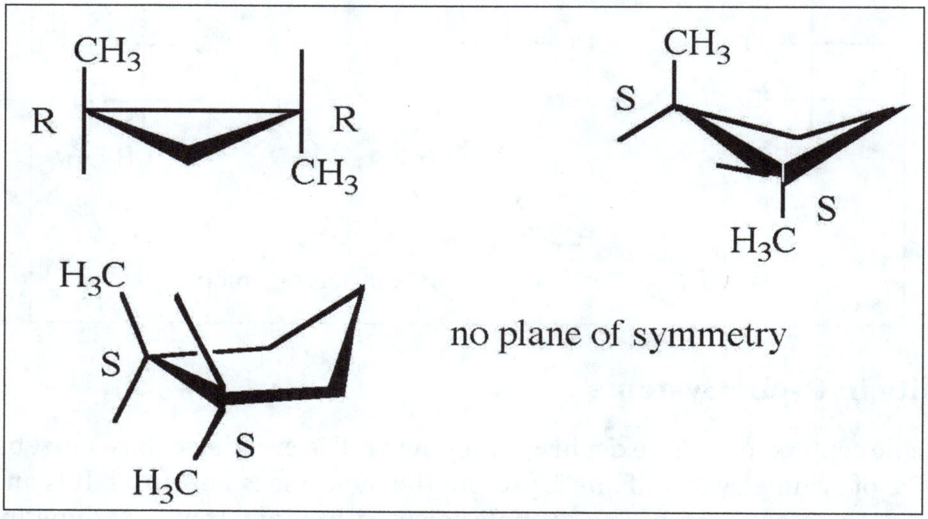

Two chiral atoms.

Chiral Systems without Chiral Centers

Ring systems do not have to contain a chiral center to be chiral. Certain molecular distortions cause ring systems to lose their symmetry and show chirality. The molecule hexahelicene, synthesized by M. S. Newman, is chiral and can be resolved into enantiomers that show optical activity. The chirality in the molecule occurs because the rings are distorted to avoid bumping into each other. The computerized structure on the right shows the shape of the molecule.

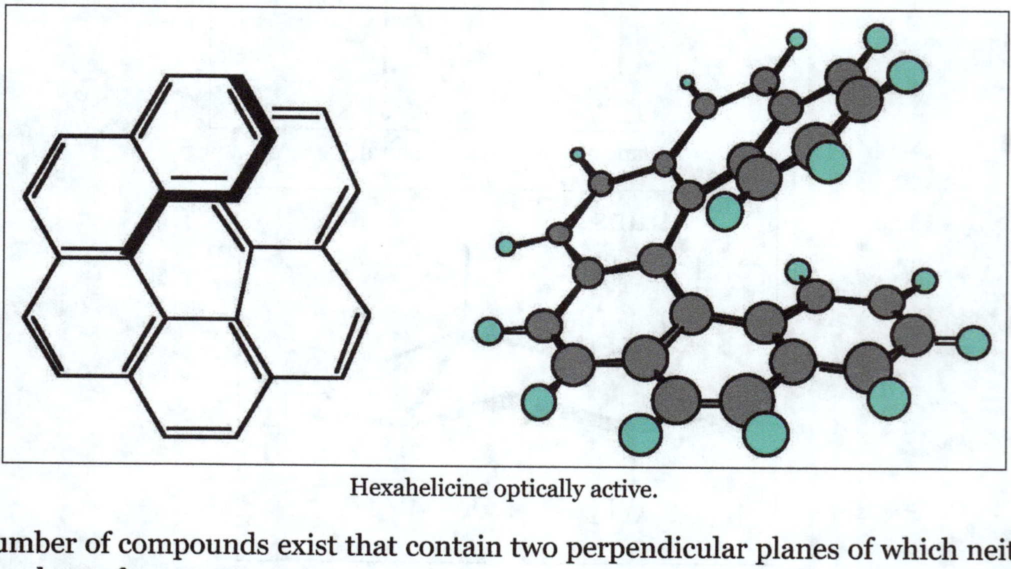

Hexahelicine optically active.

A number of compounds exist that contain two perpendicular planes of which neither has a plane of symmetery.

Several examples are shown below:

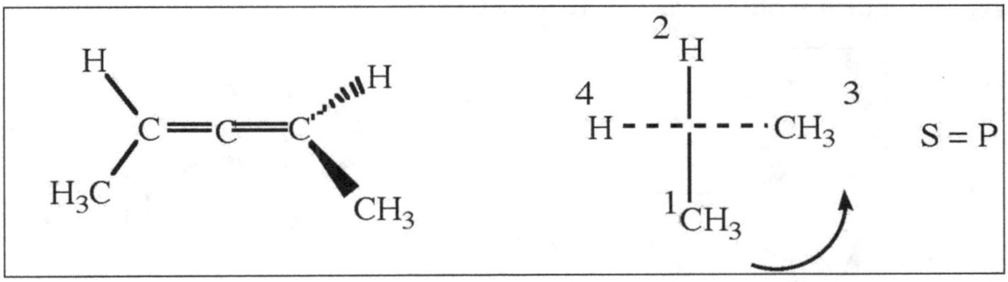

Allene

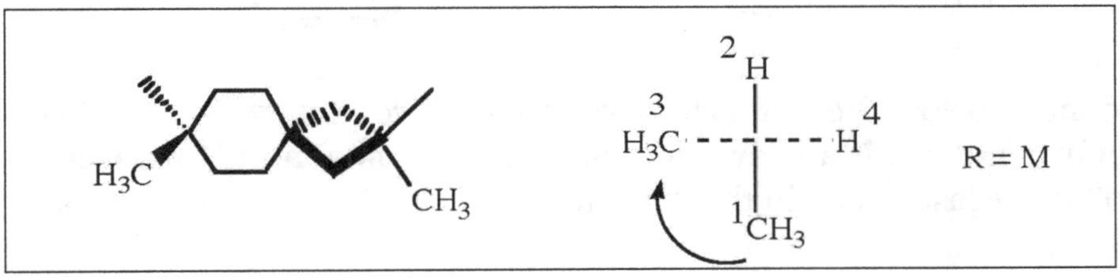

Spiro

Biraryls compounds contain perpindicular rings because the ortho substituents prevent rotation around the bond joining the rings. The compounds can be made racemic by heating and the free energy required is in the range of 10 - 50 kcal/mole depending on the substituents. The isomerization caused by restricted rotation of a carbon-carbon bond is called atromisomerism.

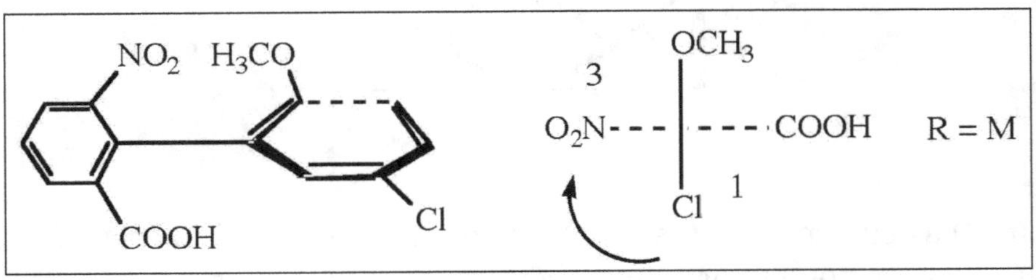

Biphenyl

Atriopisomerism is also observed in the perfluorvinyl compound below because the close proximity of the vinyl groups restricts free rotation of the groups. Chiral isomers can be isolated and have an energy difference for inter-conversion of 15.1 kcal/mol.

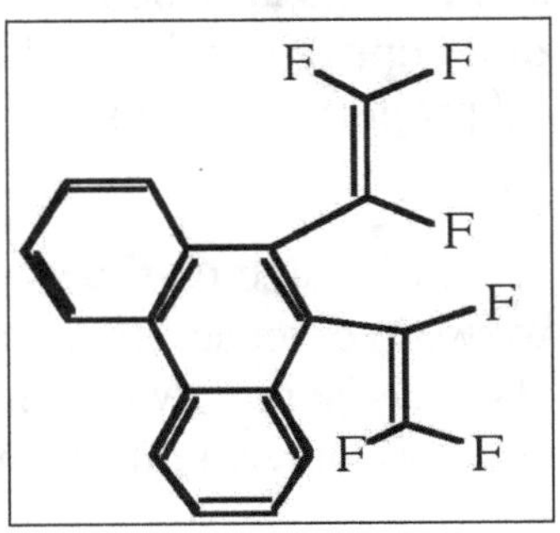

Restrictricted rotation in the compounds below also give rise to dissymetric planes of symmetry, and chiral isomers can be resolved.

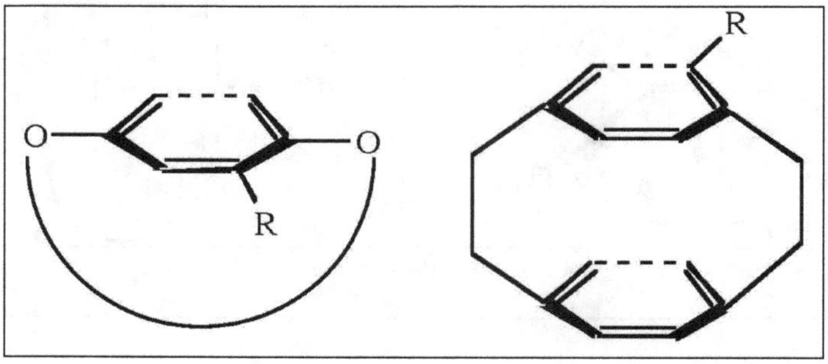

The cyclophane below shows much more restricted rotation when the R is a fluorine atom than when the R is a hydrogen atom. Ring flipping in the hydrogen compound occurs 10^{11} times faster than in the fluorine compound. This also shows that F is larger than H.

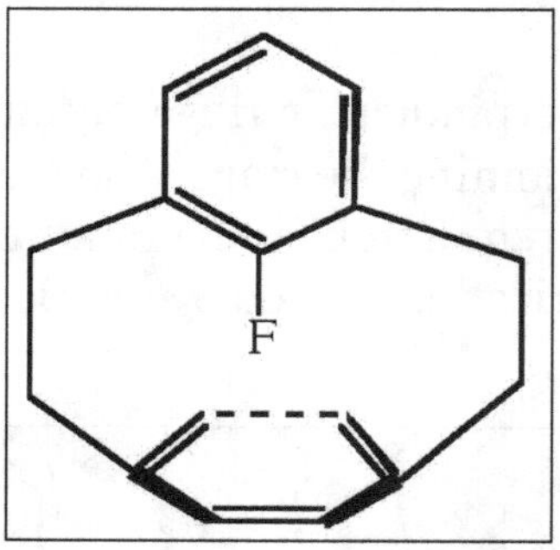

Alkenes

Alkenes are Diasterioisomers. Cis and trans or EZ isomers of alkenes are stereoisomers. Since they are not mirror images they are diasterioisomers.

Geometrical Isomers in Alkenes

The p-bond in an alkene does not permit rotation, thus all of the atoms attached directly to the alkene lie in a plane. Groups attached to the alkene could be positioned on the same side of the alkene or on opposite sides of the alkene. Such compounds are different in chemical and physical properties as well as in their geometry, and are called geometrical isomers.

In 2-butene the methyl groups can be located on the same side or on the opposite side of the double bond, giving rise to two geometrical isomers. The isomer with the methyl groups on the same side is called the cis isomer, while the isomer with the groups located on opposite sides is called the trans isomer. Trans isomers of compounds are usually more stable than cis isomers.

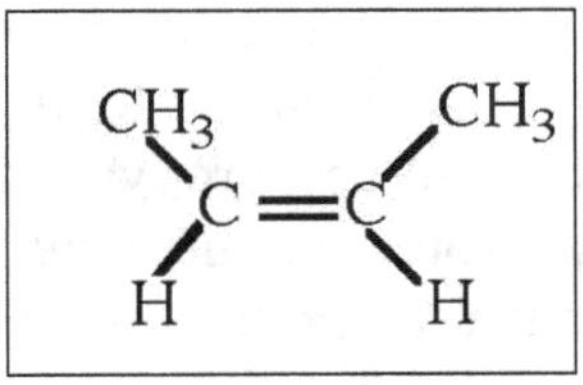

cis- 2-butene.

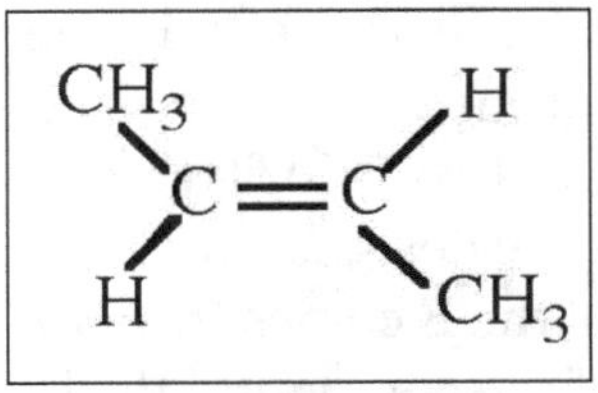

trans- 2-butene.

Except for very simple alkenes with hydrogen atoms on each carbon of the alkene, the designations of cis and trans for alkenes are replaced by a system that uses E and Z designations.

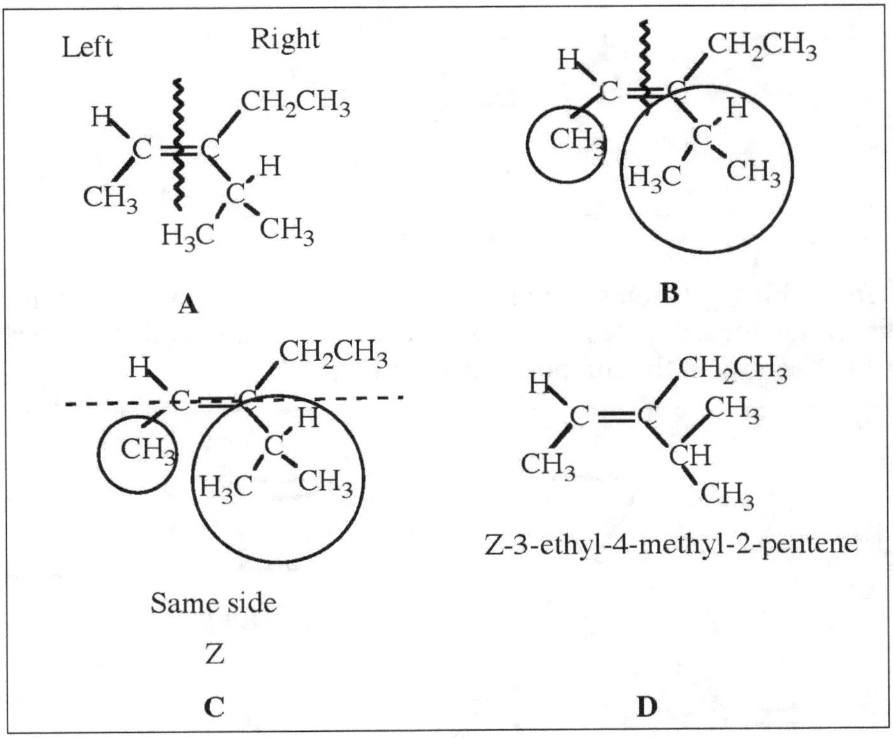

In the E-Z system the geometry is specified by the relative positions of the two highest priority substituents on the two carbons of the double bond. The priorities of the substituents are determined by the atomic number with atoms of higher atomic number having higher priority. In the example above, the molecule is divided into left and right sides. The group of higher priority on the left side is determined. The left side has a methyl and a hydrogen attached. The carbon atom of the methyl group has a higher atomic number than hydrogen and is given the higher priority (circled in B). On the right side the two groups are ethyl and isopropyl. The carbon atoms have the same atomic number but the isopropyl group has three carbons attached to the alkene carbon whereas the ethyl group has two carbons attached. Thus the isopropyl group has higher priority (circled in B).

Next in structure C a determination is made to locate the two priority groups relative to a horizontal plane. If the two groups are on the same side of the horizontal line then the isomer is designated Z. If the two priority groups were on opposite sides of the horizontal line, the isomer would be designated E. The Z comes from the German word zusammen, meaning same, while the E comes from the German word entgegen, meaning opposite. In the final name of the compound the letter designating the geometry, in this case Z, is written preceding the name of the compound (D). The structure of the E isomer is shown below.

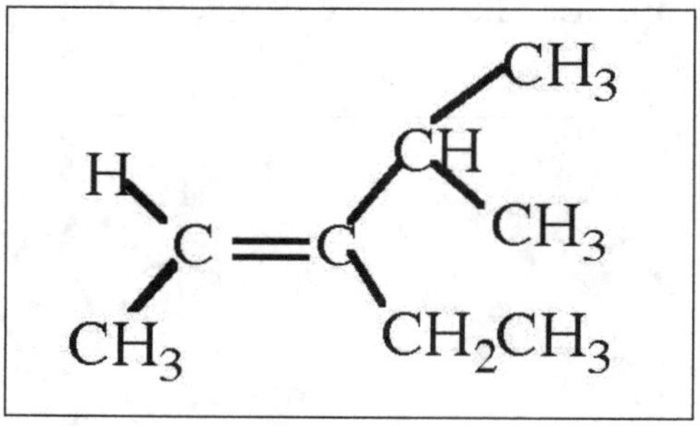

E-isomer

Cyclic alkenes with 3 - 7 atoms in the ring are fixed in the cis or Z structure. A trans or E configuration for a double bond can exist in cyclic compounds but a large ring is necessary or the strain in the compound is very high.

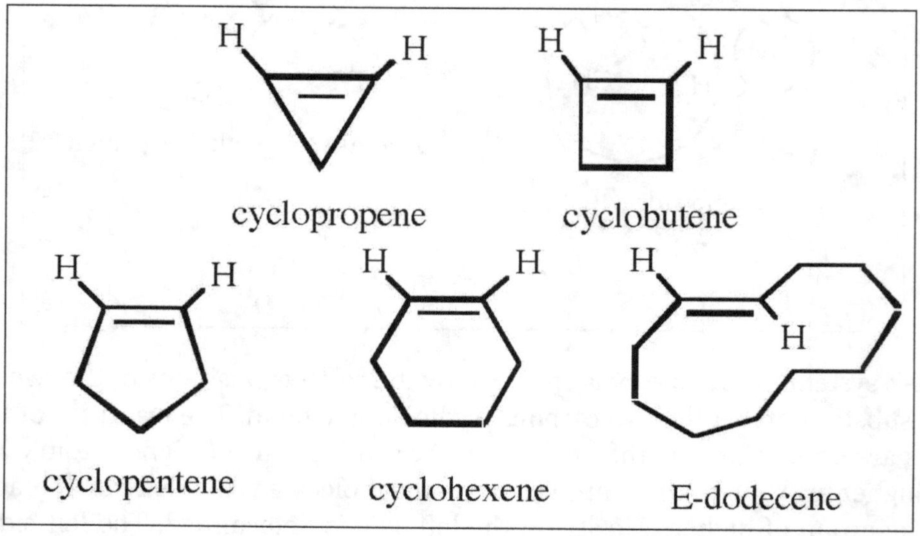

Geometric Isomers in Cyclic Systems

Substituents attached to a ring system will either be on the same side of the ring or on the opposite side of the ring. Thus, cyclic alkanes show cis and trans geometrical isomers. The letters E and Z are not used in cyclic alkanes. The cis and trans isomers of 1,3-dimethylcyclobutane are shown below.

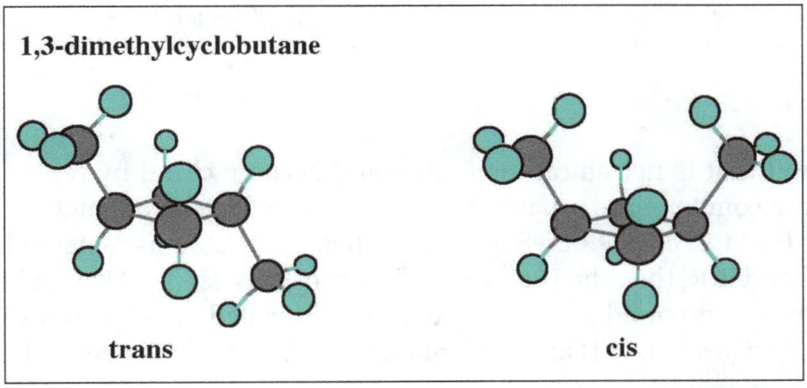

Stereospecific Reactions

Some reactions give only one stereoisomer as the product. If the reactions starting material has two stereoisomers (such as cis and trans) and each material gives only one stereoisomer as a product of the reaction, then it is called stereospecific. But both stereoisomeric starting materials must be used in the reaction. The Simmons Smith reaction shown is a stereospecific reaction.

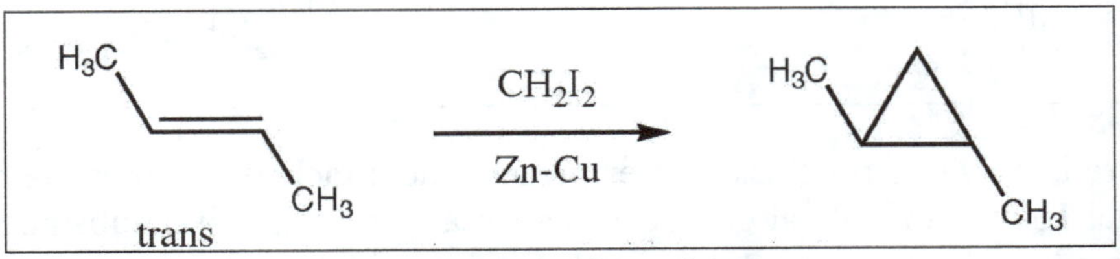

Simmons-Smith reaction.

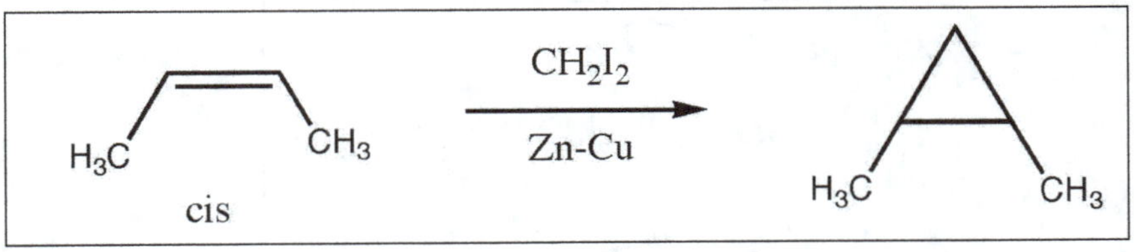

Stereoselective Reactions

A reaction that has the possibility of forming more than one stereoisomer, but actually forms one isomer more than any other is stereoselective. There does not need to be a comparison with another starting material.

This E2 reaction give more trans than cis, and is stereoselective.

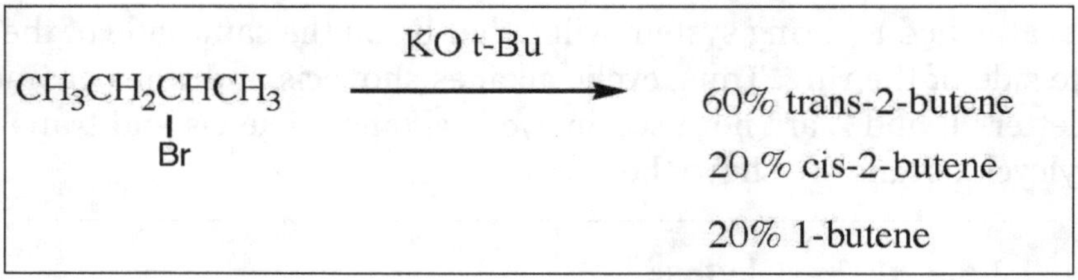

Prochirality

If a compound that is not chiral (achiral) could become chiral by replacement of an atom, then the compound is prochiral. Replacement of the H atoms in the compound below would lead to a chiral R an S mixture. When the H atom is replaced by D and the chiral compound is R, then the H atom is called pro-R. When the H atom is replaced by D and the chiral compound is S, then the H atom is called pro-S. It is a simple matter on paper to change a H to a D and determine the chirality of the new molecule.

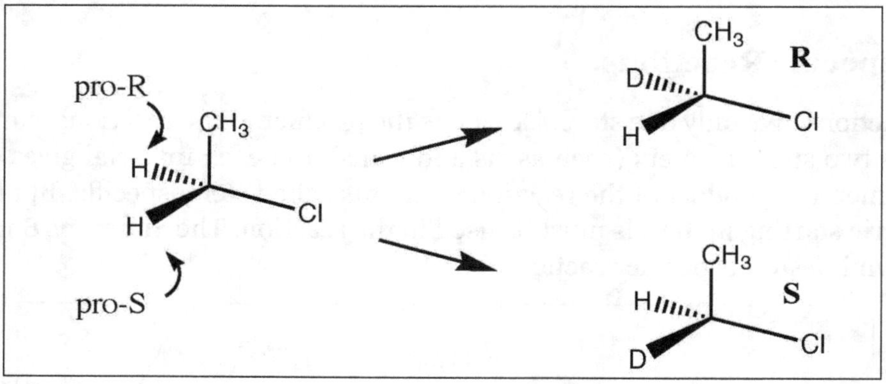

When there is only one prochiral center present the attached H atoms are called enantiotropic. They are identical except in their reaction with chiral substances. When NADH is oxidized by enzyme catalysis, only the pro R hydrogen is used.

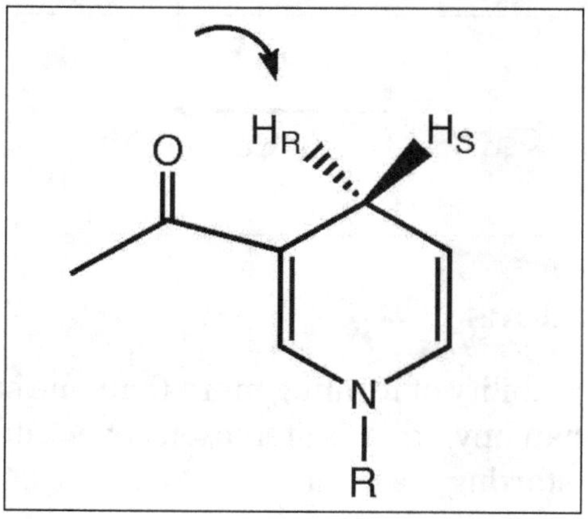

In cases where exchange of a hydrogen would lead to diasterioisomers the hydrogens are called diastereotropic. Diastereotropic hydrogens are different in many ways, but importantly is that they are not equivalent in a H NMR spectrum. Thus in the compound below the methylene protons are not equivalent and thus have different chemical shifts and undergo splitting with each other as well as with the methine hydrogen.

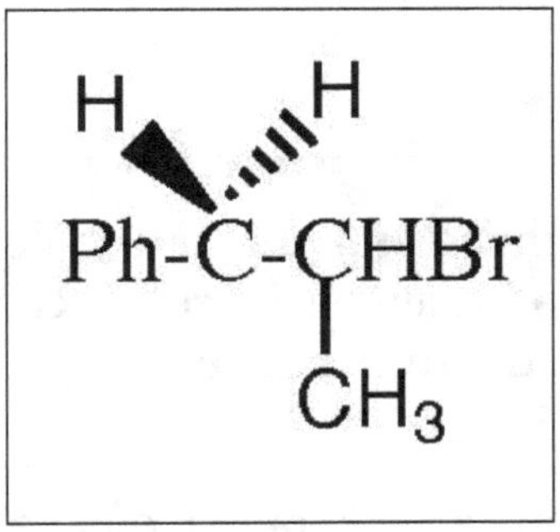

Re and Si faces of a C=O group.

Addition of certain reagent (Grignard, hydride) to a carbonyl group often gives rise to a chiral center. The carbonyl function can be designated as Re (R) or Si (S) by the same rules used for chiral centers (high priority to low priority). Then the carbonyl group is considered have a Re face and a Si face for stereochemical considerations.

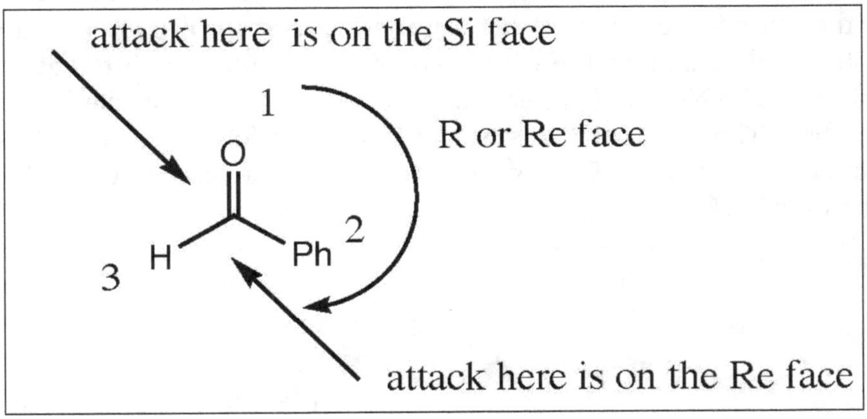

Conformational Isomers: Acyclic Systems (Non-cyclic)

Ethane

The atoms attached to an alkane carbon arrange such that they are as far apart from each other as possible. This arrangement causes a tetrahedral shape, and is shown below for methane. The length of the carbon-hydrogen bond is 1.09 Å, and all H-C-H bond angles are 109.5°. Ethane has a C-C bond length of 1.53 Å and the bond angles are all 109.5°.

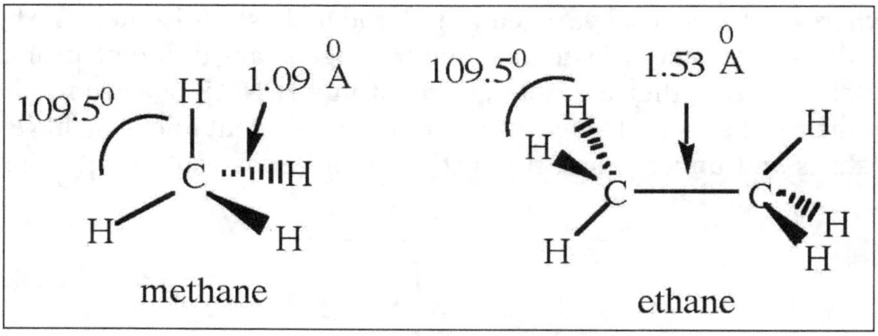

The structures shown for methane and ethane use dotted-line wedge formulae. In this representation the dark bonds are visualized as coming toward the reader, while the dotted lines go away from the reader. The straight bonds are in the plane of the page.

Free rotation around the carbon-carbon bond in ethane leads to many different structures called conformational isomers, or conformers. Two major types of structures occur during the bond rotation: one called eclipsed in which the C-H bonds are directly across from each other, and the other called staggered in which the C-H bonds are lined up between each other. The staggered conformation is more stable by 3 kcal/mol than the eclipsed one.

The conformations are shown in four different common representations; the sawhorse structure, the dotted-line wedge structure, the Newman projection and a computer generated ball and stick model. Each type of structure allows a different perspective on the molecule. The Newman projection is viewed on a line from the front carbon to the back carbon. The front carbon is not written but is located at the intersection of the three bonds to hydrogen. The big circle represents the back carbon with the three hydrogens tilting back.

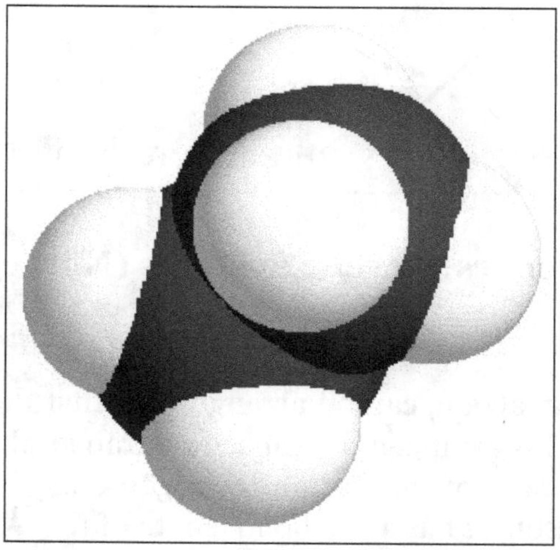

Space filling model of staggered ethane.

Eclipsed Conformation of Ethane.

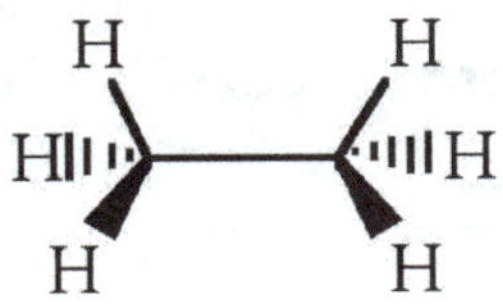 Dotted-line Wedge	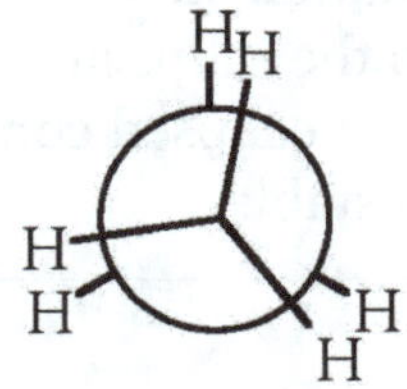 Newman
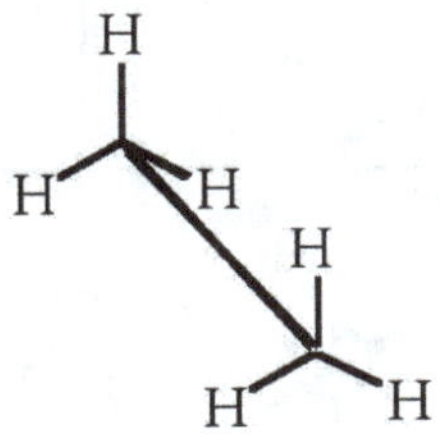 Sawhorse	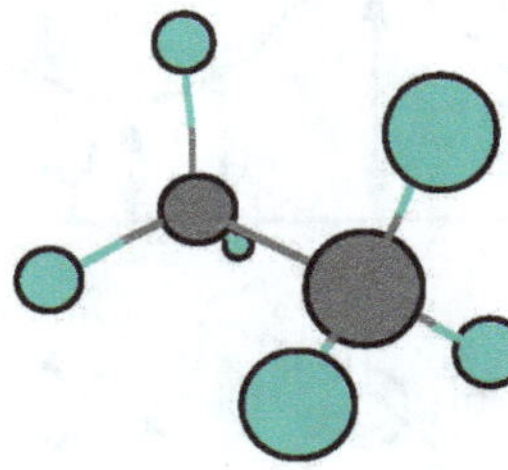 Ball and stick

Staggered Conformation of Ethane.

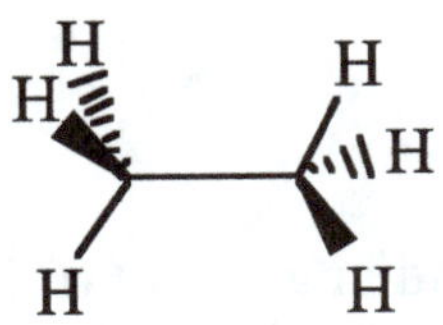 Dotted-line Wedge	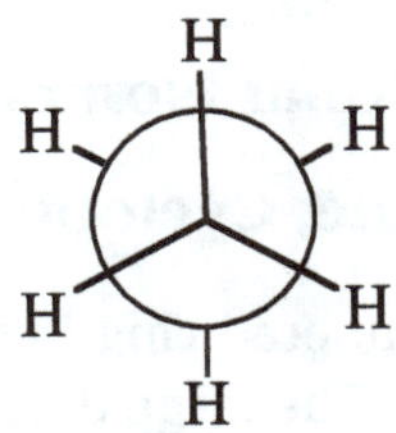 Newman
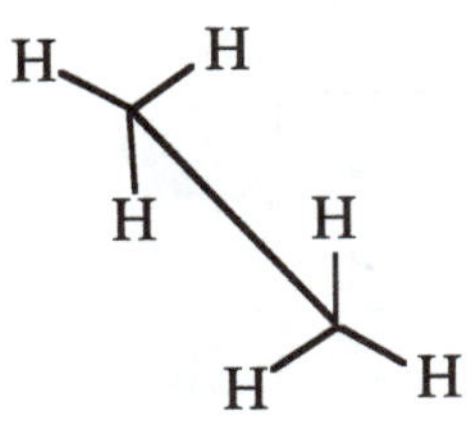 Sawhorse	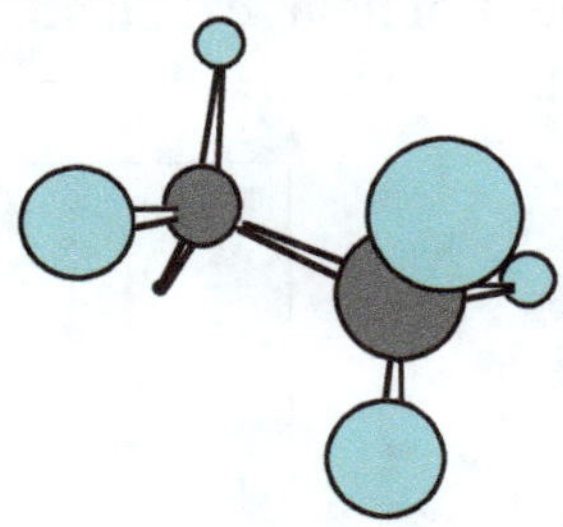 Ball and stick

Butane

Rotation by 60° of the bond between carbon two and carbon three in butane gives rise to several conformational isomers. The most stable conformation occurs in a staggered structure in which the methyls are opposite each other, called the anti-conformation. Rotation of the bond in the anti-conformation 60° leads to an eclipsed conformation. Continued rotation leads to a staggered conformation in which the methyl groups are close to each other, called the gauche conformation. Further rotation gives the least stable conformation in which the hydrogens are eclipsed and the methyls are also eclipsed, called the eclipsed methyl conformation. Staggered conformations are always more stable than eclipsed conformations, and structures with crowding of the methyl groups are less stable.

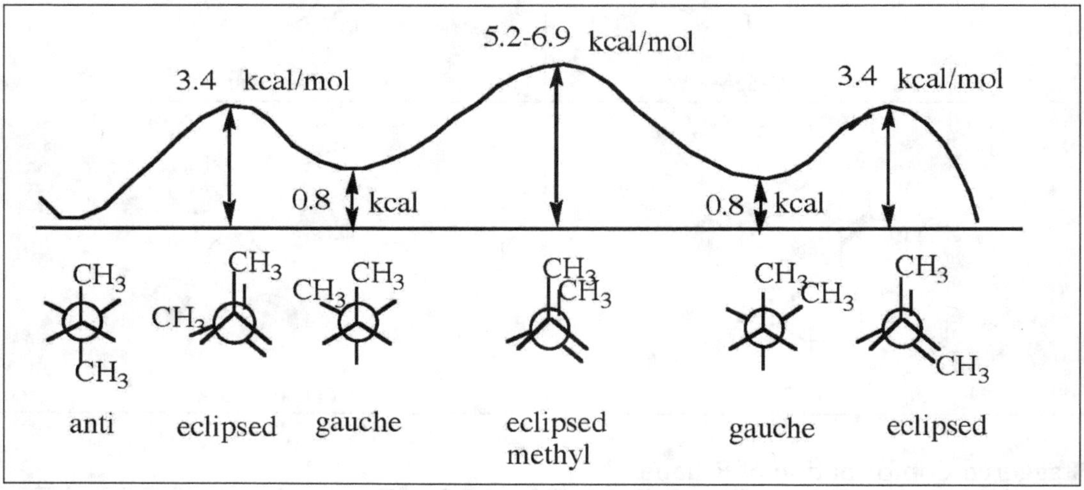

Conformational Isomers in Ring Systems

Cyclopropane, Cyclobutane and Cyclopentane

The three-membered ring in cyclopropane contains bond angles of 60° instead of the normal 109.5°. The large deformation of bond angles results in considerable strain in the molecule. On combustion, molecules with strain in their bond angles produce a higher amount of heat, called strain energy. The strain energy in cyclopropane is about 27 kcal/mol.

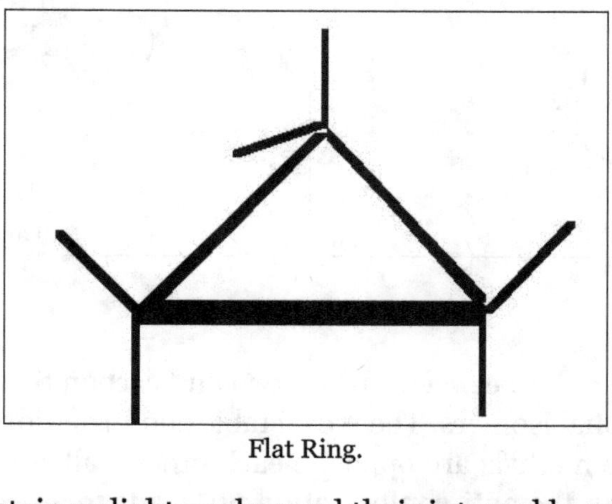

Flat Ring.

Cyclobutane rings contain a slight pucker and their internal bond angles are 90_{0}. Thus a strain energy of 26 kcal/mol exists in cyclobutane. Cyclopentane has a shape of an envelope and the internal bond angles are 108o. Thus cyclopentane does not contain significant strain energy.

Cyclobutane

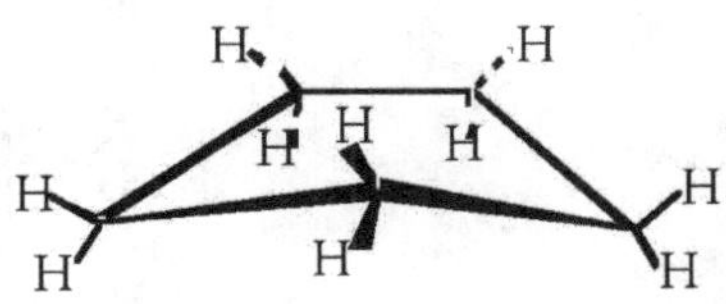

Cyclopentane

Cyclohexane

The cyclic six membered ring is the most commonly found ring system in organic chemicals. In both nature and in the laboratory, chemical reactions produce six membered rings with ease. Cyclohexane is not strained, and it contains important conformations. The most stable and most important conformation of cyclohexane is the chair conformation shown below. An unstable conformation of cyclohexane is the boat conformation which is 7.1 kcal/mol higher in energy than the chair form.

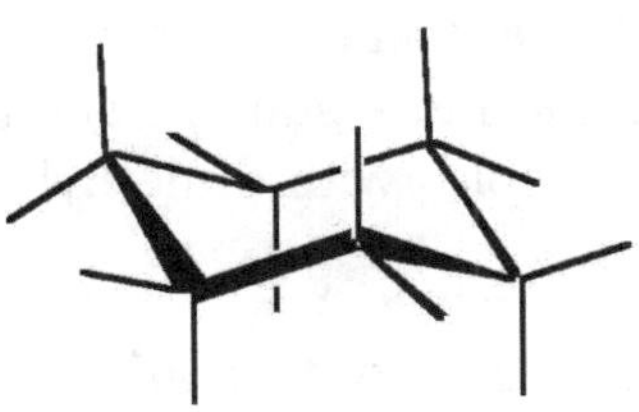

Boat

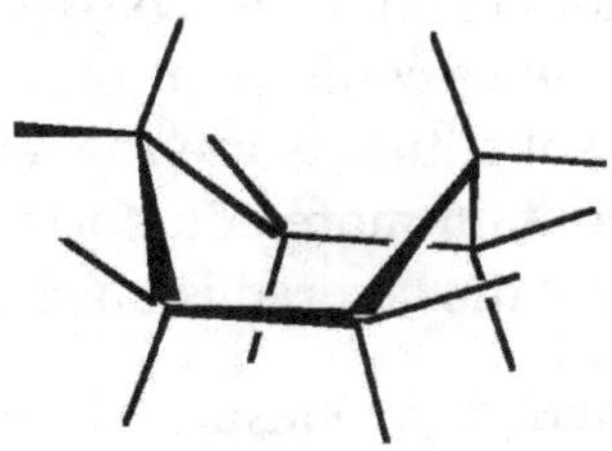

Chair

The boat and chair conformations are inter-convertable by passing through some very unstable high energy structures called the half-chair and the twist-boat conformations.

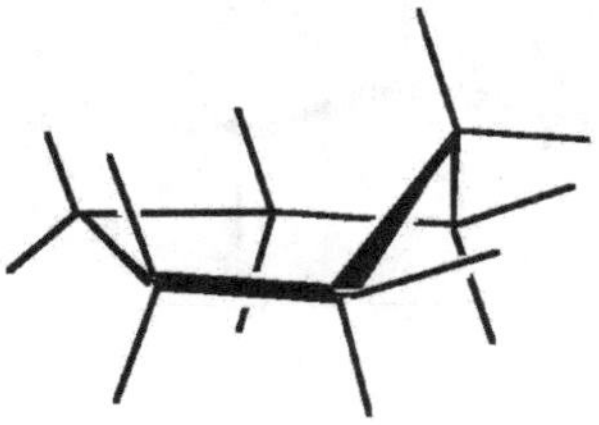

Half-chair

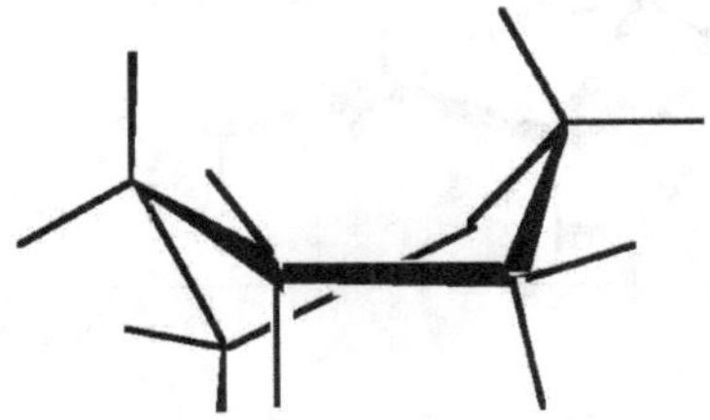

Twist-boat

The chair form of cyclohexane is flexible, and may be flipped into other chair forms. The outer bonds, called equatorial bonds, flip into vertical bonds, called axial bonds. Thus in cyclohexane the axial and equatorial bonds interchange when the molecule flips from one chair form to another. A portion of the conformational energy diagram for cyclohexane is shown below. In the completed diagram the final stable structure is another cyclohexane ring with the axial and equatorial substituents interchanged.

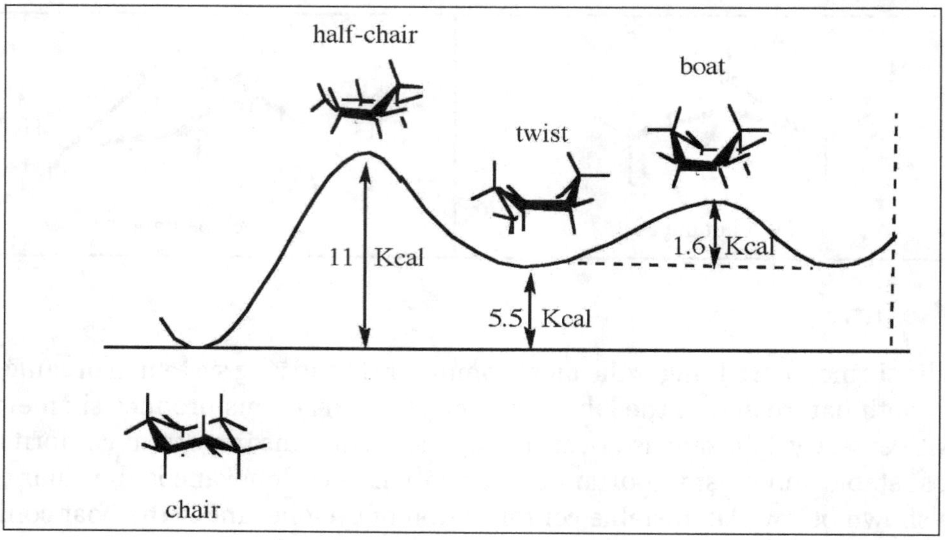

When the cyclohexane ring contains substituents, the chair forms that result from the conformational flipping can be of different energy. The more stable conformational isomer, also called a conformer, is the one usually with the least crowding of substituents. Equatorial substituents are the least crowded and the more stable structures are the ones that contain more equatorial substituents. Examples with dimethyl cyclohexane isomers show the favored isomer in the equilibrium.

Cis-1,2-dimethylcyclohexane can interconvert to another conformation, but the new conformation is identical with the starting structure because one methyl is axial while the other is equatorial in each structure. Thus there is no energy difference.

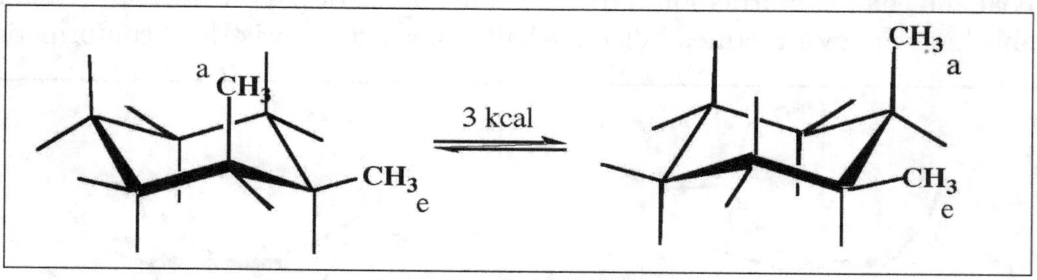

cis-1,2-dimethylcyclohexane

In trans-1,2-dimethycyclohexane one conformation exists with both methyls equatorial while the other conformer has both methyls axial. The structure with two equatorial methyls is much more stable and represents the major isomer in the equilibrium. The reason for instability in the diaxial isomer is that the methyls become crowded by the hydrogens in other axial positions, as shown by the dotted lines in the diaxial isomer. The crowding is termed a 1,3-diaxial interaction.

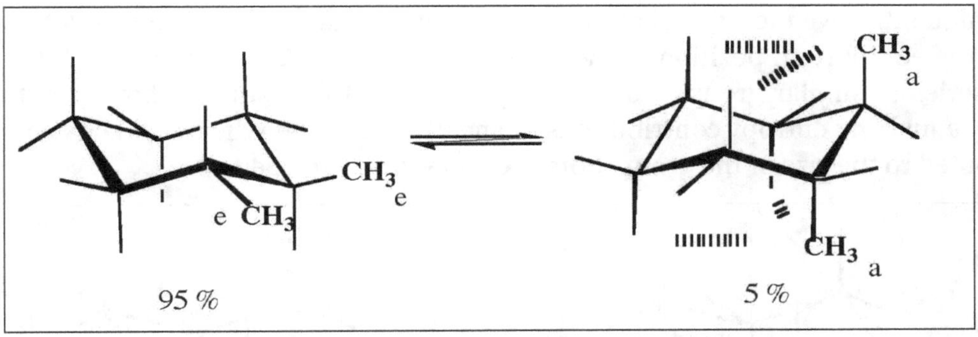

trans-1,2-dimethylcyclohexane

Several other dimethylcyclohexanes are shown below in their more stable conformation. The trans-1,3- and the cis-1,4-dimethyl compounds are the same when they interconvert to the other conformer, thus only one conformer exists.

cis-1,3-dimethylcyclohexane two equatorial methyls.	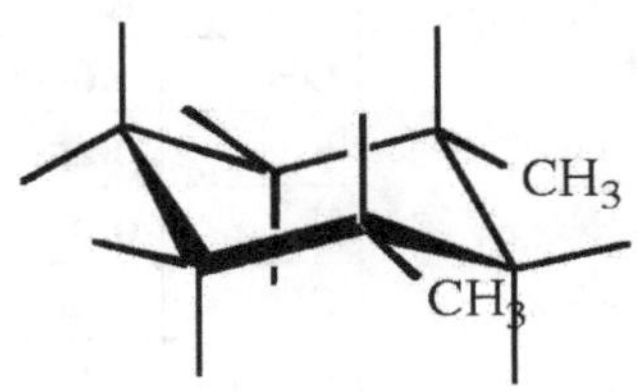
trans-1,3-dimethylcyclohexane one equatorial and one axial methyl.	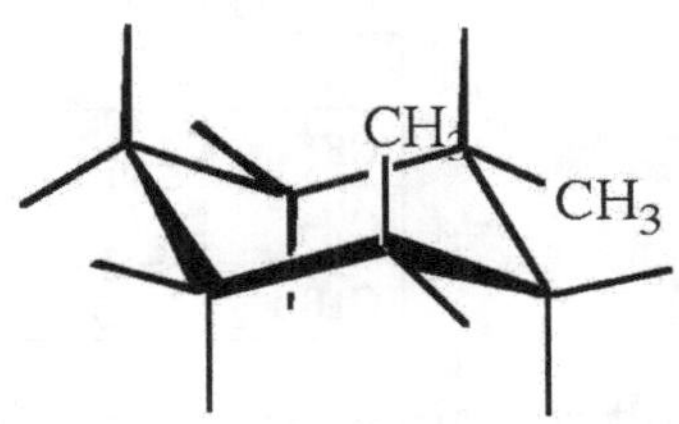
cis-1,4-dimethylcyclohexane one equatorial and one axial methyl both conformations are the same.	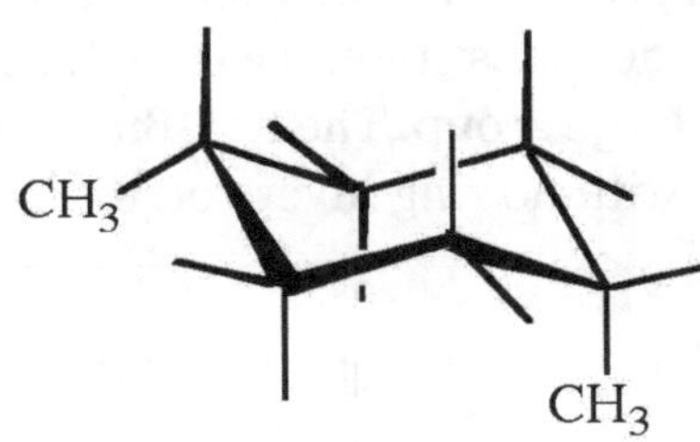
trans-1,4-dimethylcyclohexane two equatorial methyls.	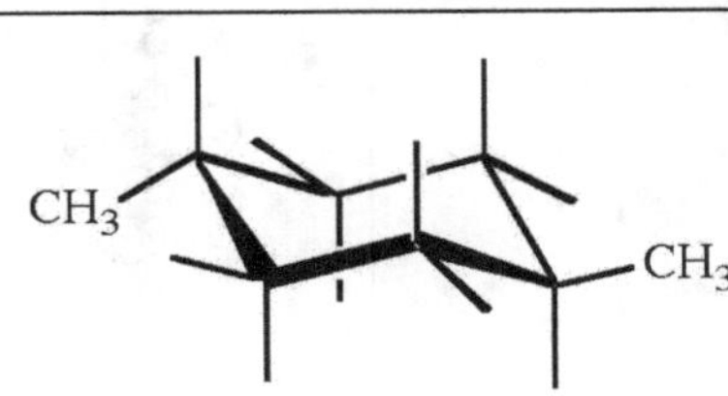

A semiquantitative measure of steric size is given by the A value. The A is determined from the equilibrium position of the anti-equitorial pair. A is set to equal - DG. Thus in a series of similar groups where the cyclohexane to substituent bond lengths are nearly equal and entropy contributions are nearly equal, then the free energy change is attributed to the size of the group. Some A values are listed below.

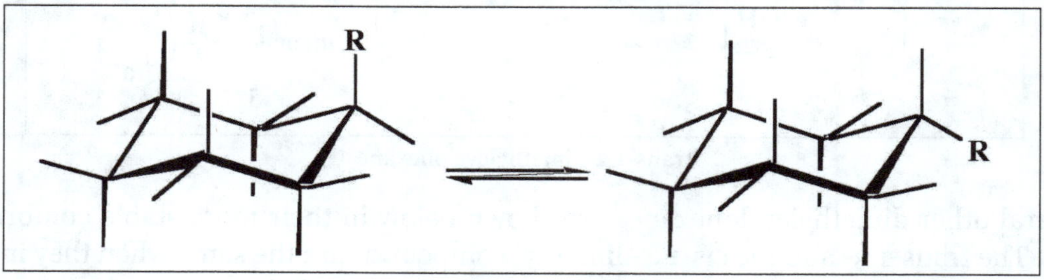

$A = -\Delta G = RT\ln K$

	A	Van der Waal	Es	v
OH	0.6-1.0	1.52	+.69	
F	0.25-0.42	1.47 A	+.78	
H	0.0	1.2 A	+1.24	
CH_3	1.74	16.8	0.0	0.52
Et	1.8		-.07	0.56
Iso-Pr	2.2		-.47	0.76
t-Bu	4.7		-1.24	1.24
CF_3	2.4	42.6	-1.16	0.91
C_6H_5	2.2			
OtBu	0.75			

These result contain a value for F that is not representative of the steric size of F. This happens because of the different C-F bond length and contributions from entropy. But for the other substituents we see that CF_3 and iso-Pr are nearly the same and that t-Bu is a very large group. The tert-Bu group is thus confined in cyclohexane to an equitorial position with no ring inversion to the axial isomer.

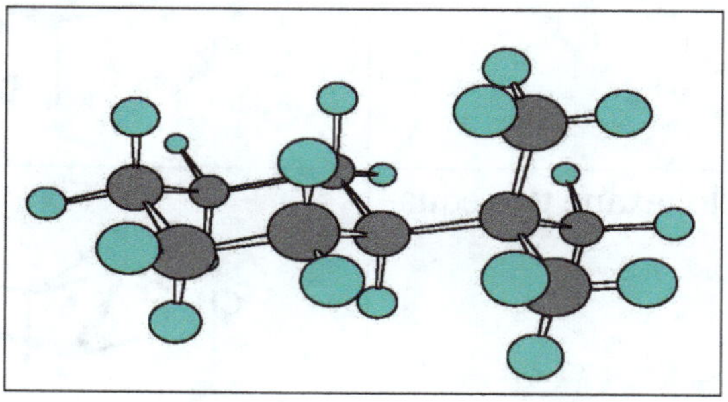

In some six-membered rings containing oxygen, an alpha substituent, if it is electro- negative such as OH, prefers that axial conformation. This is known as the Anomeric effect and is seen often in carbohydrates such as glucose. The exact reason for this effect is uncertain, but it seems that the dipoles of the electronegative atoms are partially neutralized when the substituent is axial.

Anomeric Effect		
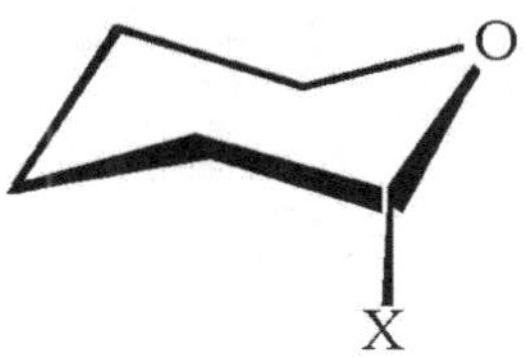	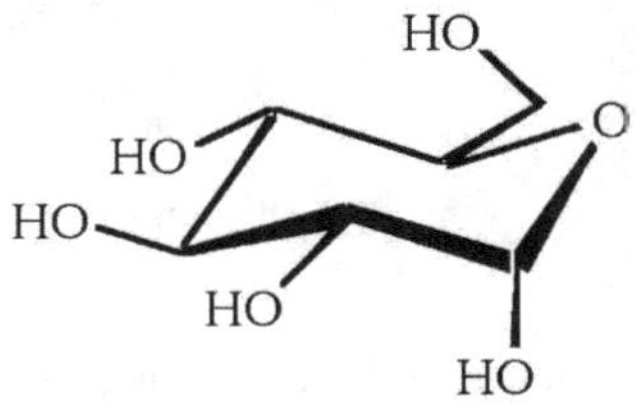	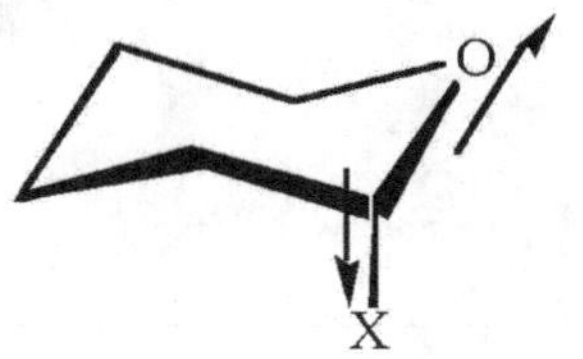
X = OH, Cl, F	Glucose	Dipoles partially cancel

Anomeric effect and is seen often in carbohydrates such as glucose. The exact reason for this effect is uncertain, but it seems that the dipoles of the electronegative atoms are partially neutralized when the substituent is axial. When two 1,3 oxygen atoms are present in a six-membered ring substituents in the 5 position often prefer the axial conformation as the 1,3-diaxial repulsions are not present.

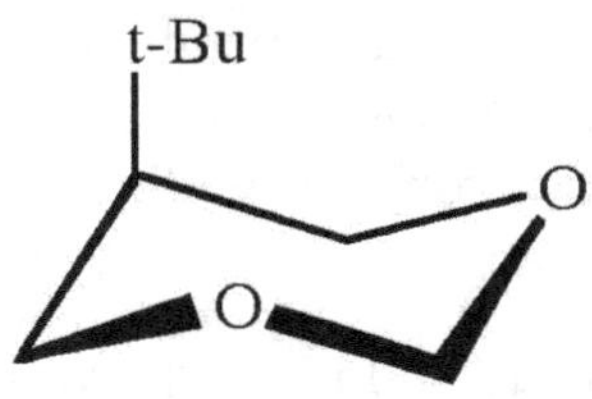

Medicinal Chemistry

Medicinal chemistry is a discipline that encloses the design, development, and synthesis of pharmaceutical drugs. The discipline combines expertise from chemistry, especially synthetic organic chemistry, pharmacology, and other biological sciences. It is also part of medicinal chemistry the evaluation of the properties of existing drugs.

The use of plants, minerals, and animal parts as medicines has been recorded since the most ancient civilizations. With the evolution of the knowledge the means for drug discovery also evolved. New molecules with potential pharmaceutical interest, "hits', are natural products, or compounds generated by computational chemistry, or compounds from a screening of chemical libraries, from combinatorial chemistry, and from pharmaceutical biotechnology. The "hit" compound is improved for its pharmacologic, pharmacodynamic and pharmacokinetic properties by chemical or functional group modifications, transforming it into a lead compound. A lead compound should have a known structure and a known mechanism of action. The lead compound is further optimized to be a drug candidate that is safe to use in human clinical trials.

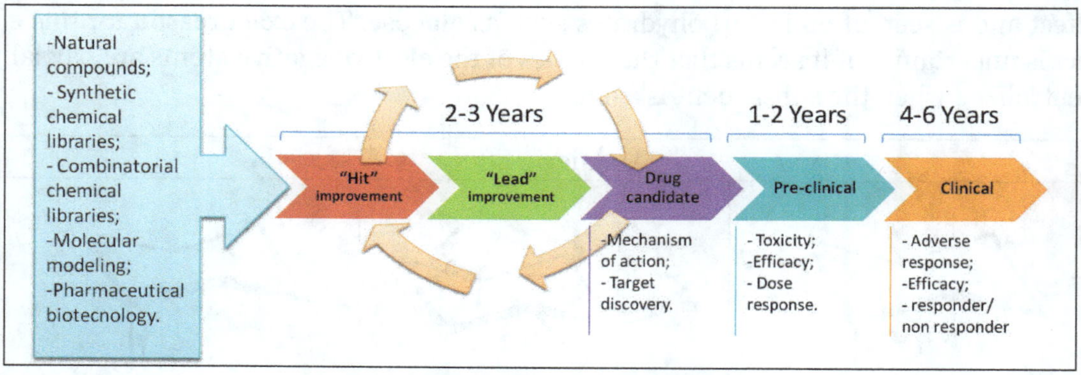

A time frame for drug discovery.

Drug Discovery

Natural Products

Natural products have been major sources of lead compounds in the discovery of new drugs for the treatment of infectious diseases, lipid disorders, neurological diseases, cardiovascular and metabolic diseases, immunological, inflammatory and related diseases, and oncologic diseases. From the pharmaceutical entities approved worldwide between 1981 and 2006, 5.7% were natural products and 27.6% were natural-product derived molecules, whereas from 2005 to 2010, seven natural products and 12 natural product derivatives were approved for use in clinical practice. Traditional medicine, however, can also produce false leads, as reviewed recently for curcumin.

Table: Examples of natural products and natural product derivatives currently in clinical use.

Disease area	Generic name(Trade name)	Lead compound	Year
Antibacterial	Doripenem (Finibax/Doribax)	Theinamycin	2005
Antiparasitic	Fumagillin (Flisint)	Fumagillin	2005
Oncology	Romidepsin (Istodax)	Romidepsin	2009
Alzeimer's disease	Galantamine (Reminyl)	Galantamine	2002
Immunosuppression	Mycophenolate sodium (Myfortic)	Mycophenolic acid	2003
Dyslipidemia	Rosuvastatin (Crestor)	Mevastatin	2003
Pain	Capsaicin	Capsaicin	2009
Diabetes	Exenatide (Byetta)	Exenatide	2006

Selection of an Organism

Natural compounds have high chemical diversity. They come from different organisms. The choice of plants, microorganisms, fungi or other organisms for investigation for new compounds tend to be based on random screening, selection of specific taxonomic groups, a chemotaxonomic group of secondary metabolites like alkaloids, database surveillance of a species collection, or ethnomedical approach. Several drugs trace their origin to ethnobotanical use.

Sample Collection

0.3 to 1 Kg of dried plant or 1 Kg of wet weight of marine samples should be collected from different parts of the organism. If the sample needs to be re-collected, it should be performed in the same localization and at the same time during the day, since different habitats can lead to different secondary metabolites. Of importance is also the deposit of a representative voucher specimen in a central repository, for later access. Besides plants, in the last decade the interest in fungi, cyanobacteria of marine origin and the evolution of the “genome mining” techniques for culturing microorganisms in the laboratory led to the discovery of new natural products.

Extraction

The extraction protocols differ from laboratory to laboratory. The initial extraction of plants usually is done with a polar solvent, like methanol or ethanol. It is then subject to a defatting process through partition with a nonpolar solvent, such as hexane or petroleum ether. The extract is partitioned between a semipolar solvent, such as chloroform or dichloromethane, and a polar aqueous solvent. Marine and aquatic organisms are extracted fresh with methanol or dichloromethane. Vegetable tannins have to be removed from the extracts. Their presence can lead to protein precipitation and enzyme inhibition, interfering with the biologic assays.

Purification and Isolation

Separation of the active compound is done using several chromatographic techniques. High pressure liquid chromatography (HPLC) and its coupling with high-throughput screening (HST) assays simplify the purification and isolation of active compounds. The compounds are identified by nuclear magnetic resonance (NMR) and mass spectrometry (MS). These techniques, when coupled with liquid chromatography (LC), allow simultaneous purification and structure elucidation of biologically active compounds. Purification and isolation of the active compound are measured by biologic assays. HTS assays can evaluate a large number of extracts or compounds in a cell-based or non-cell-based context.

In addition to HPLC, a new method combining high-speed counter-current chromatography (HSCCC) and size exclusion chromatography with a Sephadex LH-20 has been effectively applied to extract neuroprotective compounds from the marine brown algae, Ecklonia maxima. The structure of the isolated compounds was verified by NMR and MS.

The major challenge in natural compounds development is to obtain significant amounts for further development. Re-collection of the species of origin, or plant tissue culture, or cultivation and fermentation in large scale for microbes of terrestrial origin can produce the compounds in large scale. The natural compounds can also be synthesized. For example, Kawano S et al achieved the total synthesis of one of halichondrins, a group of structurally complex natural products isolated from various marine sponges, and addressed the issue of limited material supply.

Computational Chemistry

Molecular modeling, or more generally, computational chemistry, has become a well-established part of drug development. Molecular modeling searches new molecules, based on a theoretical platform or by screening a library. Crystallographic and/or NMR information on receptors and specific targets, such as that of cystic fibrosis transmembrane conductance regulator (CFTR), allows through molecular modeling techniques, the design of new molecules for the target. Other approaches use known active molecules as a target, and design new molecules or search the libraries for similar molecules.

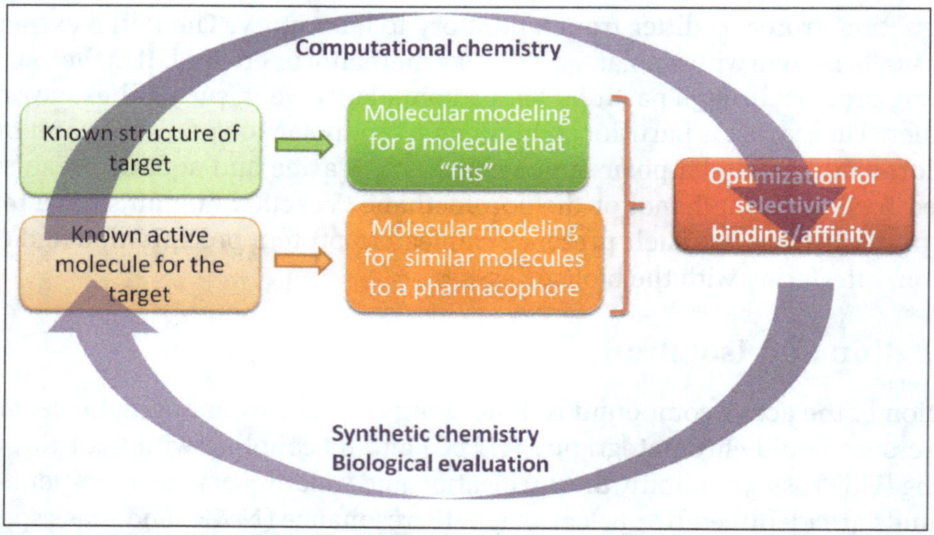

Computational chemistry approaches for the development of new drugs.

Modeling of protein-protein interactions (PPI) is becoming a valuable tool for the development of new therapeutic strategies targeting selective PPI modifiers. For instance, a new method combines cross-linking mass spectrometry with modeling approaches to generate antibody inhibitors of the interplay between R7BP, a regulator of itching sensation, and RGS7/Gβ5. Inhibitory activity was evaluated by surface plasmon resonance spectroscopy.

The found "hit" molecules can either be synthesized in the laboratory or purchased. After evaluation and structure-activity-relationship (SAR) disclosure, new studies of molecular modeling can be performed to find a more active molecule or to optimize the found molecule.

Computational chemistry is based on the visualization and manipulation of three-dimensional molecular models. Rotation of bonds, structure building, molecular mechanisms and/or dynamics, conformational analysis, electronic properties, molecular surface displays and calculation of various physical properties are possible through molecular modeling. The techniques used are molecular mechanisms and molecular dynamics simulations, Monte Carlo techniques, ligand docking, and virtual screening methods.

The current state-of-the art systems allow working with more than 20 molecules and thousands of molecular surface points in real time. Each molecule can be color labeled and controlled in three dimensions, where the intra-molecular distances and the non-contiguous dihedral angles can be adjusted and monitored. Shape, charge and hydrophobicity of the atoms in the molecule can also be simulated. The electrostatic potential gradient or electrical field can also be displayed graphically using short vectors. Molecular modeling requires make-bond, break-bond, fuse rings, delete-atom, add-atom, add-hydrogens, invert chiral center, etc. These operations should allow a refined structure for a selected target.

A fragment-based reinforcement learning method has been implemented to design new compounds with specific functions. This method, deep fragment-based multi-parameter optimization, uses long short-term memory networks to create new molecules with required properties from lead compounds by modifying their parts. An actor-critic model, the base of this approach, implements temporal difference learning.

Several systems were developed for storing and retrieving the information generated by molecular modeling. The number of new molecules generated by molecular modeling can number thousands for a target, making the synthesis and biological evaluation of all these new molecules a challenge. The development of virtual screenings allowed overcoming the previous problem. After virtual screening evaluation only the molecules with the required biological activity would be synthesized or purchased for further biological evaluation.

2

Organic Compounds and their Classification

Organic compounds are the chemical compounds which contain carbon-hydrogen bonds. Organic compounds are classified into acyclic and cyclic compounds. Cyclic organic compounds are further divided into homocyclic and heterocyclic compounds. The topics elaborated in this chapter will help in gaining a better perspective about the classification of organic compounds.

Organic Compounds

The compounds in solid, liquid or gaseous state which contain carbon in its molecule are known as organic compounds. There are a large number of organic compounds and therefore a proper systematic classification was required. Organic compounds can be broadly classified as acyclic (open chain) or cyclic (closed chain). Moving on to their classification in detail:

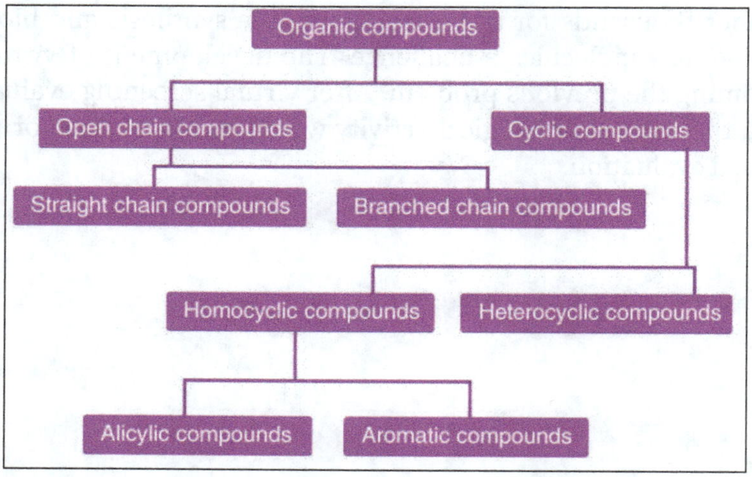

Organic compounds are seen in a number of formats, including Lewis structures, space filled models and structural formulas. It is not uncommon to view the hydrogens as lines or to leave them all together in a structural formula of an organic molecule. They are understood to be present in order to complete the 4-bonds provided by the carbon atoms.

Organic compounds have been detected by mass spectra. The results obtained indicate that all extractants were able to perform direct analysis after extraction without any clean-up phase. Nevertheless, the form of organic compounds extracted depended on the solvent used, suggesting their distinct ability to solubilize various biosolid organic compounds.

Organic chemistry was once thought to be confined to the study of substances produced as part of the natural processes of living organisms, but as Friedrich Wohler discovered in the early 1800s, organic compounds can be synthesized from minerals and other non-organic materials in the laboratory. Indeed, modern chemistry and materials sciences have concentrated on the remarkable properties of carbon atoms for the production of synthetic chemicals, pesticides and a host of other things. Organic compounds contain carbon, almost always bonded to another carbon and/or hydrogen.

Sometimes, other elements, such as phosphorus, nitrogen and oxygen, are also bound to carbons. There are a few carbon compounds that are not considered organic molecules. Those involve carbon dioxide, carbon monoxide, cyanates, cyanides and other carbon-containing ion compounds. Other elements, such as phosphorus, nitrogen and oxygen, are also sometimes bound to carbon. There are a few carbon compounds which are not considered organic molecules. Those involve carbon dioxide, carbon monoxide, cyanates, cyanides and other carbon-containing ion compounds.

Alcohols include chemicals such as ethanol and isopropanol. They are used as antiseptics and ethanol is a staple of the beverage industry. Finally, carboxylic acids include a wide range of chemicals, including pharmaceuticals. Aspirin, one of the oldest commercial medicines, contains carboxylic acid.

While there are millions of organic compounds, there is a fairly simple classification scheme for these compounds and a method for naming even the most complex organic compounds. Organic compounds also contain carbon, along with other elements essential for the reproduction of living organisms. Carbon is the main factor as it has four electrons that can accommodate eight electrons in an outer shell. As a result, several forms of bonds can be formed with other carbon atoms and elements such as hydrogen, oxygen, and nitrogen. Hydrocarbons and proteins are strong examples of organic molecules capable of producing long chains and complex structures.

The organic compounds made up of these molecules form the basis for chemical reactions in plant and animal cells – reactions that provide the energy required to find food, to replicate, and for all other life-related processes.

Acyclic Compounds

Acyclic Compounds are compounds of the fatty series, or aliphatic compounds; organic substances (hydrocarbons and their derivatives) whose molecules do not contain rings but rather are “open” chains.

In the compounds with so-called normal structure (for example, n-butane $CH_3—CH_2—CH_2—CH_3$) the atoms of the carbon skeleton of the molecule are arranged linearly. The arrangement of the carbon atoms in the form of branching chains is called the iso-structure—for example, isobutyl alcohol:

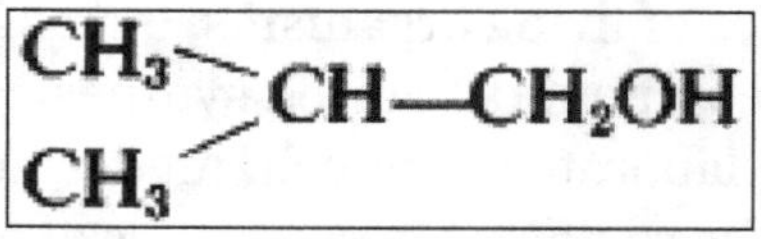

The length of the chains in acyclic compounds is extremely variable—for example, a hydrocarbon has been obtained with a linear chain containing 100 atoms of carbon: $CH_3(CH_2)_{98}CH_3$.

Acyclic and cyclic compounds are connected to each other by mutual transitions. Thus, from acetylene $CH \equiv CH$ under certain conditions it is possible to obtain acyclic vinyl acetylene $CH_2 \equiv CH—C\equiv CH$, and under different conditions the cyclic hydrocarbon benzene C_6H_6 is obtained. The transformation of saturated into aromatic acyclic hydrocarbons such as benzene and its homologues has great practical significance. Many methods for opening rings are also known. Thus, ethylene glycol is formed from ethylene oxide under the influence of water:

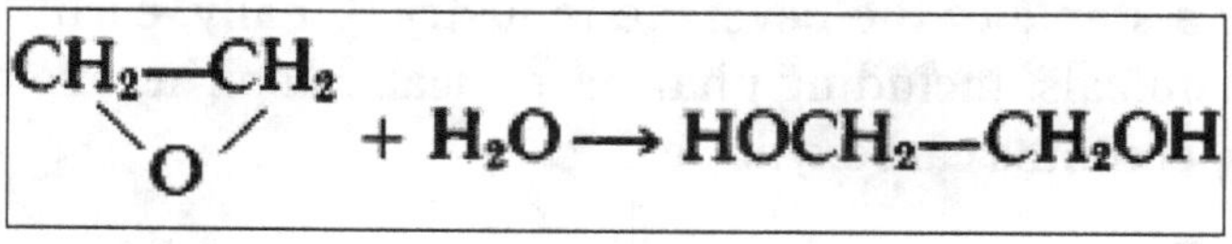

Many different classes of compounds (alcohols, ketones, acids, amines, and others) which are derivatives of hydrocarbons with open chains—saturated as well as unsaturated—belong to the acyclic compounds.

The principal source for obtaining acyclic compounds is petroleum and petroleum products. Ethylene and its homologues, acetylene, and carbon monoxide are the raw material for large-scale industrial production of many acyclic compounds. For example, synthetic rubber, polymers, ethyl and isopropyl alcohols, and acetone are obtained from olefins; from acetylene—acetic acid and chloroprene rubber; from carbon monoxide and hydrogen—methanol and synthetic gasoline; and from carbon monoxide and acetylene—acrylic acid.

Cyclic Compounds

Cyclic compound is class of compounds, mainly organic compounds, containing one or more rings (cycles) consisting of three or more atoms.

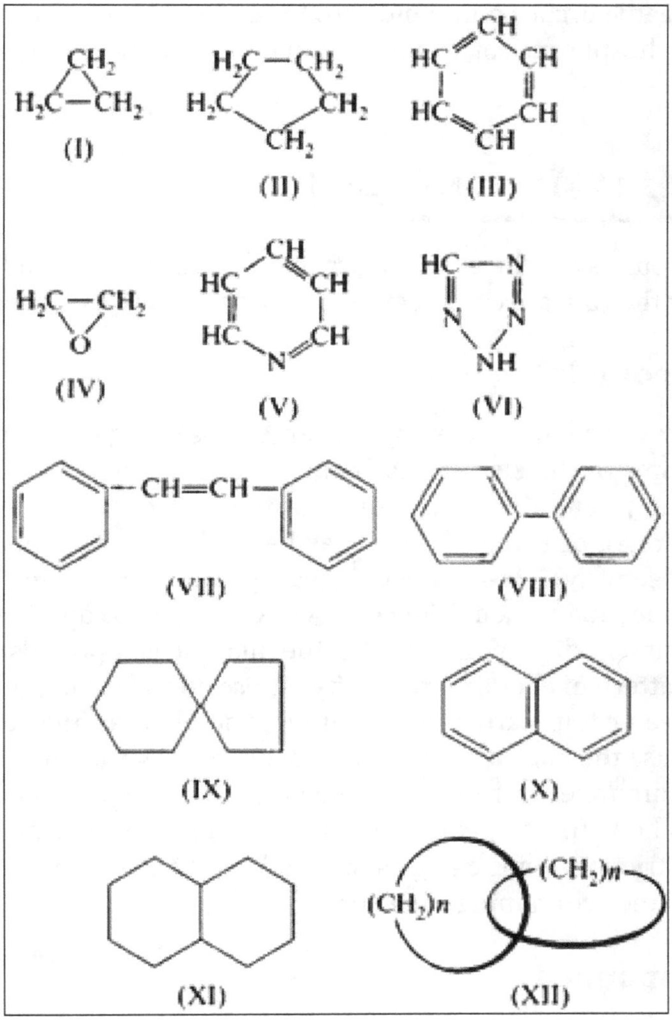

Five- and six-member rings occur most often because of the ease of their formation and the absence of ring strain. Depending on the nature of the atoms composing the ring, a distinction is made between carbocyclic, or isocyclic, compounds and heterocyclic compounds. In carbocyclic compounds, the rings consist only of carbon atoms, as in for example, alicyclic compounds, such as cyclopropane (I) and cyclopentane (II), and aromatic compounds, such as benzene (III). In heterocyclic compounds, the rings contain, in addition to carbon atoms, one or more atoms of other elements, called heteroatoms, mainly, oxygen, nitrogen, and sulfur; examples of such compounds are ethylene oxide (IV), pyridine (V), and tetrazole (VI). In polycyclic compounds, the rings may be separated from each other, as in stilbene (VII), may be attached by a single bond, as in diphenyl (VIII), may have a single common atom, or may have two common atoms (fused-ring compounds), as in naphthalene (X) and deca-hydronaphthalene (XI). Catenanes (XII) are a special case of cyclic compounds; their molecules are constructed in the manner of a simple interlinked chain.

Examples of inorganic cyclic compounds are borazole (a boron compound), phosphonitrile chlorides (phosphorus compounds), and cyclosiloxanes (silicon compounds).

Carbocyclic Compounds

Carbocyclic compounds are the group of organic chemical compounds in which all the atoms composing the ring are carbon atoms, as benzene or cyclopropane.

Alicyclic Compounds

In Chemistry, Alicyclic compound is any of a large class of organic compounds in which three or more atoms of the element carbon are linked together in a ring. The bonds between pairs of adjacent atoms may all be of the type designated single bonds (involving two electrons), or some of them may be double or triple bonds (with four or six electrons, respectively); six-membered rings for which a system of alternating single and double bonds may be envisioned, however, belong to another important class (aromatic compounds), distinguished from the alicyclic compounds by a characteristically different pattern of chemical reactivity. Those alicyclic compounds in which the ring contains three or four carbon atoms are less stable than the compounds having larger rings, because the angles formed by adjacent covalent bonds are smaller than is necessary for maximum effectiveness. In the larger rings all the bond angles have the preferred value (about 109.5°); consequently, the atoms in the ring do not lie in one plane. Similar restrictions on the angles in double and triple bonds affect the stability of alicyclic compounds containing such bonds.

Aromatic Compounds

Aromatic compounds are chemical compounds that consist of conjugated planar ring systems accompanied by delocalized pi-electron clouds in place of individual alternating double and single bonds.

They are also called aromatics or arenes. The best examples are toluene and benzene. Aromatics require satisfying Huckel's rule. Plants and micro-organisms have an exclusive route to benzene-ring compounds. The great majority of aromatic compounds in nature, therefore, are produced by plants and micro-organisms, and animals are dependant upon plants for many aromatic compounds either directly or indirectly.

Aromatic Compounds Examples

Aromatic hydrocarbon, are hydrocarbons containing sigma bonds and delocalized pi electrons between carbon atoms in a ring. For example, benzene. They are known as aromatic due to their pleasant smell.

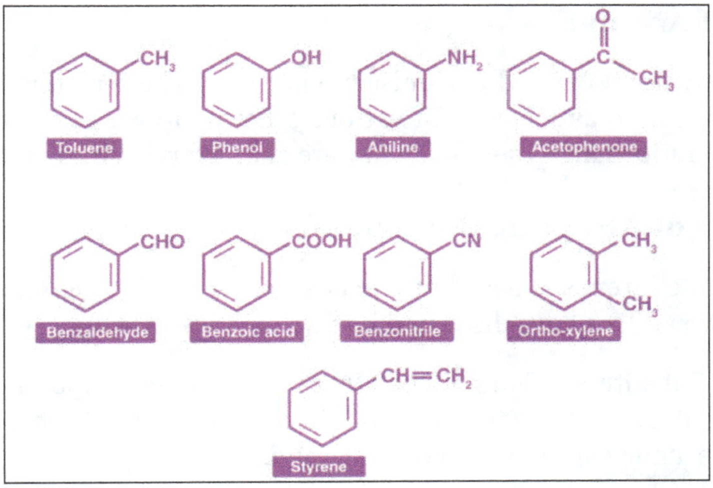

Aromatic compounds are broadly divided into two categories: benzenoids (one containing benzene ring) and non-benzenoids (those not containing a benzene ring) for example, furan. Any hydrocarbon can be classified as an aromatic compound provided they follow the Huckel rule. According to Huckel rule, for a ring to be aromatic it should have the following properties:

- Planarity.
- Complete delocalization of the π electrons in the ring.
- Presence of $(4n + 2)$ π electrons in the ring where n is an integer ($n = 0, 1, 2,...$).

Huckel's Rule of Aromaticity

Huckel's rule states that only planar, fully conjugated monocyclic polyenes having $4n + 2$ π electrons, where n is an integer, that is, $n = 0, 1, 2, 3, 4$, etc., should possess aromatic stability. An aromatic compound must be planar and contain a cyclic cloud of π electrons below and above the plane of the molecule. It contains SP^2 hybridized carbon atoms and must obey the Huckel rule.

According to this rule, the ring system must have $(4n+2)$ π electrons, where n is any whole number (0, 1, 2, 3, etc). On this basis the ring systems which have 2(n=0), 6(n=1), 10(n=2), 14(n=3) etc pi electrons are aromatic. Typical examples of aromatic compounds are benzene, naphthalene, and anthracene.

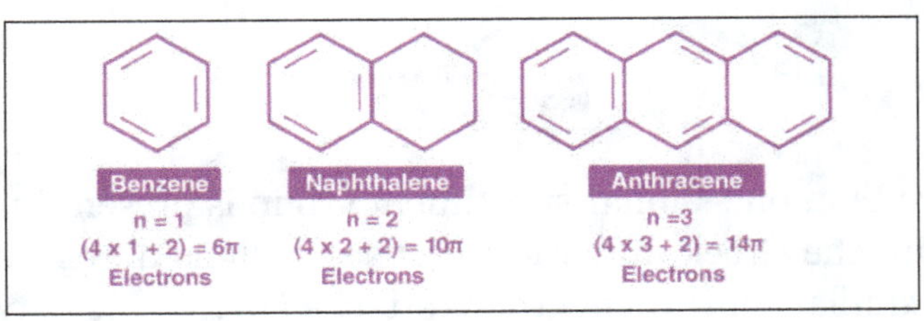

Properties of Aromatic Compounds

Arenes are mostly nonpolar and non-miscible in water. These compounds are usually unreactive and are used as solvents for various other nonpolar compounds. Their carbon to hydrogen ratio is high therefore, they are characterized by sooty yellow flame.

Classification of Aromatic Compounds

The classification of arenes is based on the position of the functional group. They are classified into two and we have discussed below:

- Nuclear Substituted Compounds: In any aromatic compound whenever any substituent or functional group, is directly linked to the benzene ring, it is known as a nuclear-substituted compound.
- Side chain Substituted Compounds: In any aromatic compound if the functional group is available in the side chain of the ring then it is known as a side chain substituted compound. These compounds are named as the phenyl derivatives of the relative aliphatic compounds.

IUPAC Nomenclature of Aromatic Compounds

Earlier, most of the compounds with the same structural formula were known by different names depending on the regions where they were synthesized. This naming system was very trivial since it raised a lot of confusion. Finally, a common naming system enlisting standard rules was set up by IUPAC (International Union for Pure and Applied Chemistry) for the naming of compounds. This method of naming is IUPAC naming or IUPAC nomenclature. IUPAC nomenclature of aromatic hydrocarbons is explained below:

- According to IUPAC nomenclature of substituted aromatic compounds, the substituent name is placed as a prefix to the name of aromatic compounds. For example, a benzene ring attached to a one-nitro group is named as nitrobenzene.

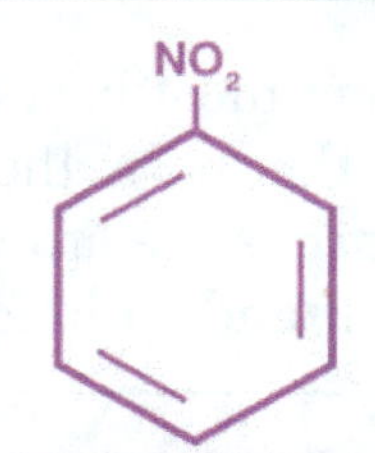

Nitrobenzene

- When more than one similar substituent group is present in the ring, they are labelled with the Greek numerical prefixes such as di, tri, tetra to denote the number of similar substituent groups attached to the ring. If two bromo- groups are attached to the adjacent carbon atoms of the benzene ring, it is named 1,2-dibromobenzene.

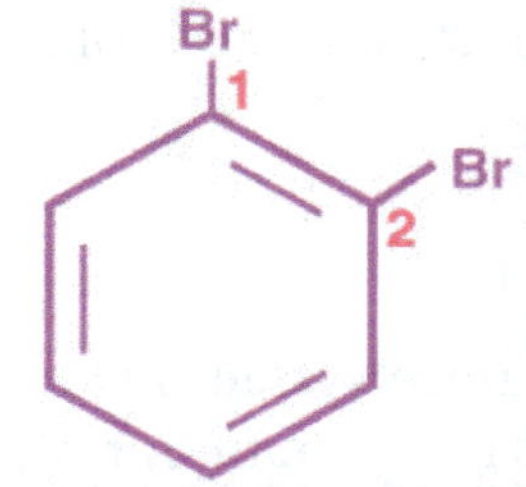

1,2 Dibromo benzene

- When different substituted groups are attached to the aromatic compounds, the substituent of the base compound is assigned number one and then the direction of numbering is chosen such that the next substituent gets the lowest number. Substituents are named in alphabetical order. For example: when chloro and nitro groups are attached to the benzene ring, we first locate the chloro group then nitro groups.

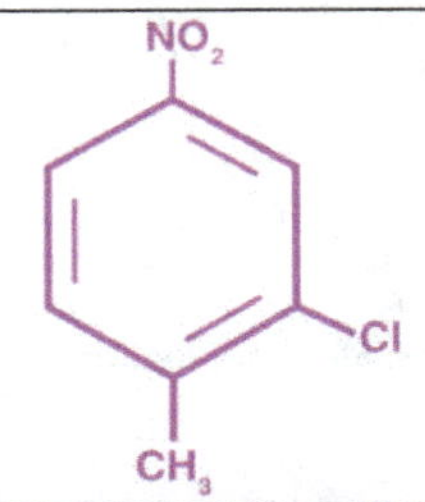

2Chloro 1-methyl 4-nitrobenzene

- In the case of multiple substituted aromatic compounds, sometimes terms like ortho (o), meta (m) and para (p) are also used as prefixes to indicate the relative positions 1,2-; 1,3- and 1,4- respectively. For example, 1,2 di-bromo-benzene can be named as o-di-bromo-benzene.

- When an alkane with a functional group is attached to an aromatic compound, the aromatic compound is considered as a substituent, instead of a parent. For example: when a benzene ring is attached to an alkane with a functional group, it is considered as a substituent named phenyl, denoted by Ph-.

Heterocyclic Compounds

Compounds classified as heterocyclic probably constitute the largest and most varied family of organic compounds. After all, every carbocyclic compound, regardless of structure and functionality, may in principle be converted into a collection of heterocyclic analogs by replacing one or more of the ring carbon atoms with a different element. Even if we restrict our consideration to oxygen, nitrogen and sulfur (the most common heterocyclic elements), the permutations and combinations of such a replacement are numerous.

Nomenclature

Devising a systematic nomenclature system for heterocyclic compounds presented a formidable challenge, which has not been uniformly concluded. Many heterocycles, especially amines, were identified early on, and received trivial names which are still preferred. Some monocyclic compounds of this kind are shown in the following chart, with the common (trivial) name in bold and a systematic name based on the Hantzsch-Widman system given beneath it in blue. The rules for using this system will be given later. For most students, learning these common names will provide an adequate nomenclature background.

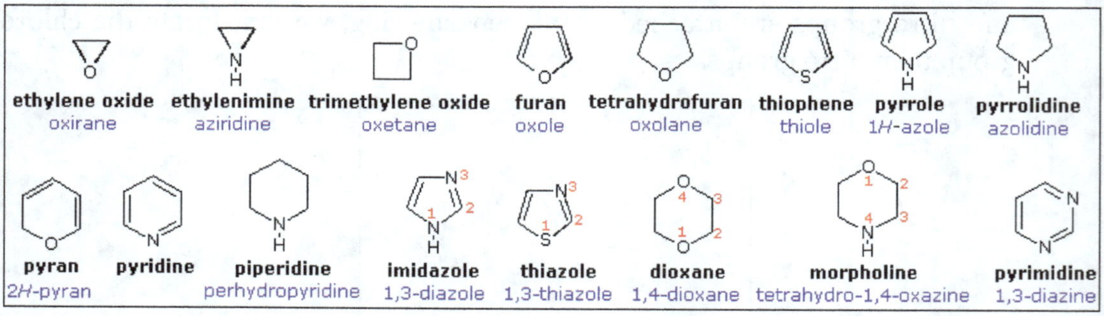

An easy to remember, but limited, nomenclature system makes use of an elemental prefix for the heteroatom followed by the appropriate carbocyclic name. A short list of some common prefixes is given in the following table, priority order increasing from right to left. Examples of this nomenclature are: ethylene oxide = oxacyclopropane, furan = oxacyclopenta-2,4-diene, pyridine = azabenzene, and morpholine = 1-oxa-4-azacyclohexane.

Element	oxygen	sulfur	selenium	nitrogen	phosphorous	silicon	boron
Valence	II	II	II	III	III	IV	III
Prefix	Oxa	Thia	Selena	Aza	Phospha	Sila	Bora

The Hantzsch-Widman system provides a more systematic method of naming heterocyclic compounds that is not dependent on prior carbocyclic names. It makes use of the same hetero atom prefix defined above (dropping the final "a"), followed by a suffix designating ring size and saturation. As outlined in the following table, each suffix consists of a ring size root (blue) and an ending intended to designate the degree of unsaturation in the ring. In this respect, it is important to recognize that the saturated suffix applies only to completely saturated ring systems, and the unsaturated suffix applies to rings incorporating the maximum number of non-cumulated double bonds. Systems having a lesser degree of unsaturation require an appropriate prefix, such as "dihydro"or "tetrahydro".

Ring Size	3	4	5	6	7	8	9	10
Suffix Unsaturated Saturated	irene irane	ete etane	ole olane	ine inane	epine epane	ocine ocane	onine onane	ecine ecane

Despite the general systematic structure of the Hantzsch-Widman system, several exceptions and modifications have been incorporated to accommodate conflicts with prior usage. Some examples are:

- The terminal "e" in the suffix is optional though recommended.
- Saturated 3, 4 & 5-membered nitrogen heterocycles should use respectively the traditional "iridine", "etidine" & "olidine" suffix.
- Unsaturated nitrogen 3-membered heterocycles may use the traditional "irine" suffix.
- Consistent use of "etine" and "oline" as a suffix for 4 & 5-membered unsaturated heterocycles is prevented by their former use for similar sized nitrogen heterocycles.
- Established use of oxine, azine and silane for other compounds or functions prohibits their use for pyran, pyridine and silacyclohexane respectively.

Examples of these nomenclature rules are written in blue, both in the previous diagram and that shown below. Note that when a maximally unsaturated ring includes a saturated atom, its location may be designated by a "#H" prefix to avoid ambiguity, as in pyran and pyrrole above and several examples below. When numbering a ring with more than one heteroatom, the highest priority atom is #1 and continues in the direction that gives the next priority atom the lowest number.

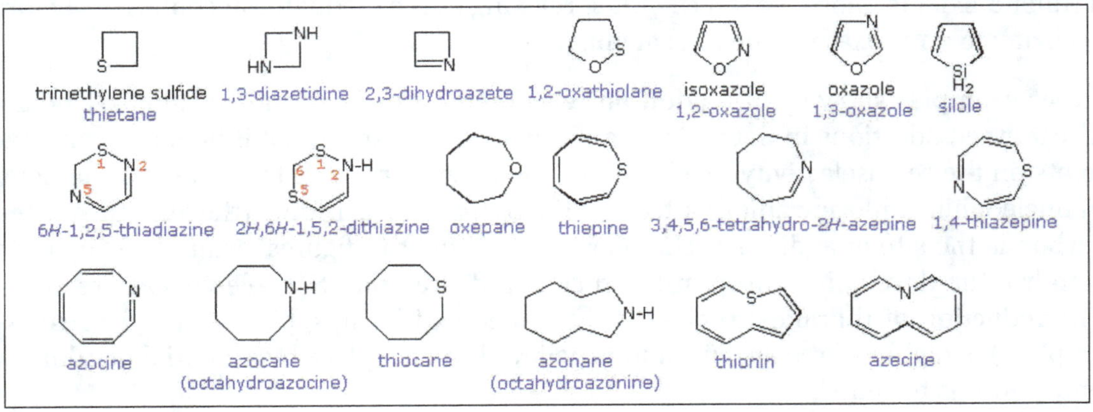

All the previous examples have been monocyclic compounds. Polycyclic compounds incorporating one or more heterocyclic rings are well known. A few of these are shown in the following diagram. As before, common names are in black and systematic names in blue. The two quinolines illustrate another nuance of heterocyclic nomenclature. Thus, the location of a fused ring may be indicated by a lowercase letter which designates the edge of the heterocyclic ring involved in the fusion, as shown by the pyridine ring in the green shaded box.

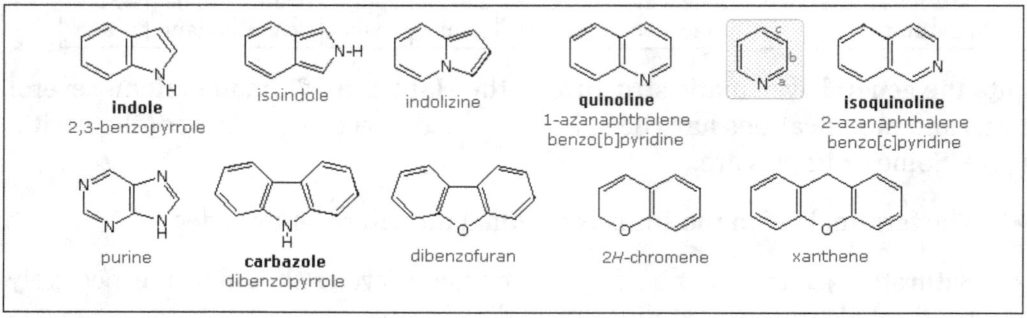

Heterocyclic rings are found in many naturally occurring compounds. Most notably, they compose the core structures of mono and polysaccharides, and the four DNA bases that establish the genetic code.

Preparation and Reactions

Three-Membered Rings

Oxiranes (epoxides) are the most commonly encountered three-membered heterocycles. Epoxides are easily prepared by reaction of alkenes with peracids, usually with good stereospecificity. Because of the high angle strain of the three-membered ring, epoxides are more reactive that unstrained ethers. Addition reactions proceeding by electrophilic or nucleophilic opening of the ring constitute the most general reaction class. Figures' example 1 in the following diagram shows one such transformation, which is interesting due to subsequent conversion of the addition intermediate into the corresponding thiirane. The initial ring opening is stereoelectronically directed in a trans-diaxial fashion, the intermediate relaxing to the diequatorial conformer before cyclizing to a 1,3-oxathiolane intermediate.

Other examples show similar addition reactions to thiiranes and aziridines. The acid-catalyzed additions in figures' examples 2 and 3, illustrate the influence of substituents on the regioselectivity of addition. Figures' example 2 reflects the S_N2 character of nucleophile (chloride anion) attack on the protonated aziridine (the less substituted carbon is the site of addition). The phenyl substituent in figures' example 3 serves to stabilize the developing carbocation to such a degree that S_N1 selectivity is realized. The reduction of thiiranes to alkenes by reaction with phosphite esters (figures' example 6) is highly stereospecific, and is believed to take place by an initial bonding of phosphorous to sulfur.

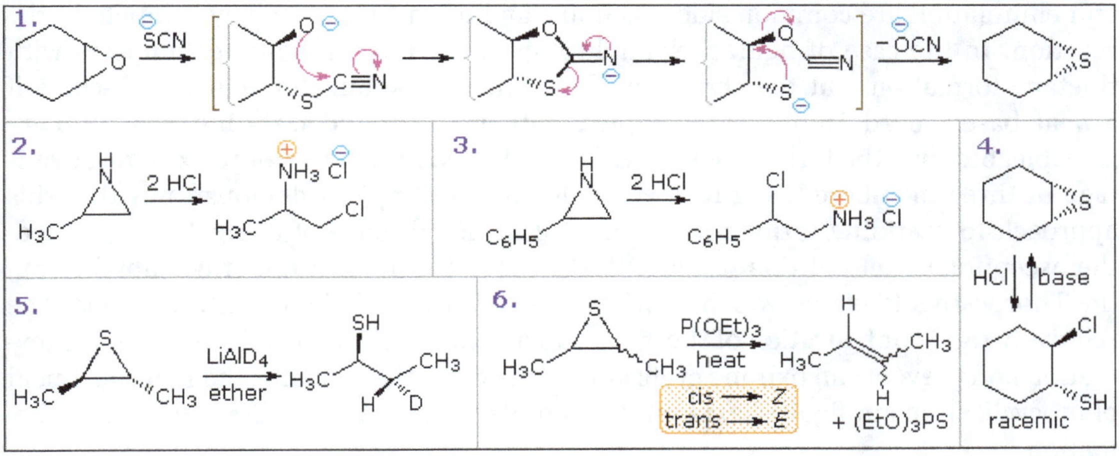

Figures' examples 7 and 8 are thermal reactions in which both the heteroatom and the strained ring are important factors. The α-lactone intermediate shown in the solvolysis of optically active 2-bromopropanoic acid (figures' example 9) accounts both for the 1st-order kinetics of this reaction and the retention of configuration in the product. Note that two inversions of configuration at C-2 result in overall retention. Many examples of intramolecular interactions, such as figures' example 10, have been documented.

An interesting regioselectivity in the intramolecular ring-opening reactions of disubstituted epoxides having a pendant γ-hydroxy substituent has been noted. Acid and base-catalyzed reactions normally proceed by 5-exo-substitution (figures' example 1), yielding a tetrahydrofuran product. However, if the oxirane has an unsaturated substituent (vinyl or phenyl), the acid-catalyzed opening occurs at the allylic (or benzylic) carbon (figures' example 2) in a 6-endo fashion. The π-electron system of the substituent assists development of positive charge at the adjacent oxirane carbon, directing nucleophilic attack to that site.

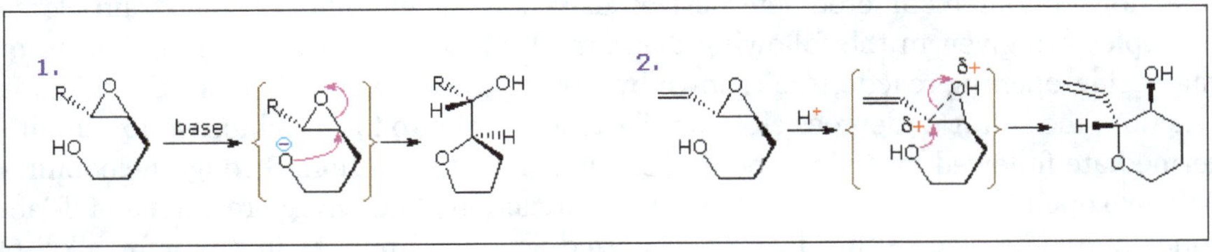

Four-Membered Rings

Preparation

Several methods of preparing four-membered heterocyclic compounds are shown in the following diagram. The simple procedure of treating a 3-halo alcohol, thiol or amine with base is generally effective, but the yields are often mediocre. Dimerization and elimination are common side reactions, and other functions may compete in the reaction. In the case of figures' example 1, cyclization to an oxirane competes with thietane formation, but the greater nucleophilicity of sulfur dominates, especially if a weak base is used. In figures' example 2 both aziridine and azetidine formation are possible, but only the former is observed. This is a good example of the kinetic advantage of three-membered ring formation. Figures' example 4 demonstrates that this approach to azetidine formation works well in the absence of competition. Indeed, the exceptional yield of this product is attributed to the gem-dimethyl substitution, the Thorpe-Ingold effect, which is believed to favor coiled chain conformations. The relatively rigid configuration of the substrate in figures' example 3, favors oxetane formation and prevents an oxirane cyclization from occurring. Finally, the Paterno-Buchi photocyclizations in figures' examples 5 and 6 are particularly suited to oxetane formation.

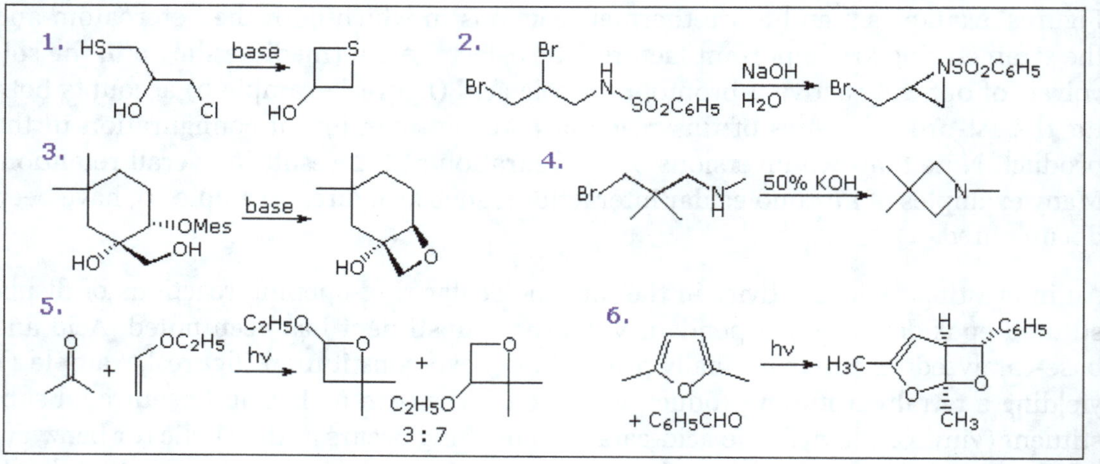

Reactions

Reactions of four-membered heterocycles also show the influence of ring strain. Some examples are given in the following diagram. Acid-catalysis is a common feature of many ring-opening reactions, as shown by figures' examples 1, 2 & 3a. In the thietane reaction, the sulfur undergoes electrophilic chlorination to form a chlorosulfonium intermediate followed by a ring-opening chloride ion substitution. Strong nucleophiles will also open the strained ether, as shown by reaction 3b. Cleavage reactions of β-lactones may take place either by acid-catalyzed acyl exchange, as in 4a, or by alkyl-O rupture by nucleophiles, as in 4b. Figures' example 5 is an interesting case of intramolecular rearrangement to an ortho-ester. Finally, the β-lactam cleavage of penicillin G (reaction 6) testifies to the enhanced acylating reactivity of this fused ring system. Most amides are extremely unreactive acylation reagents, thanks to stabilization by p-π resonance. Such electron pair delocalization is diminished in the penicillins, leaving the nitrogen with a pyramidal configuration and the carbonyl function more reactive toward nucleophiles.

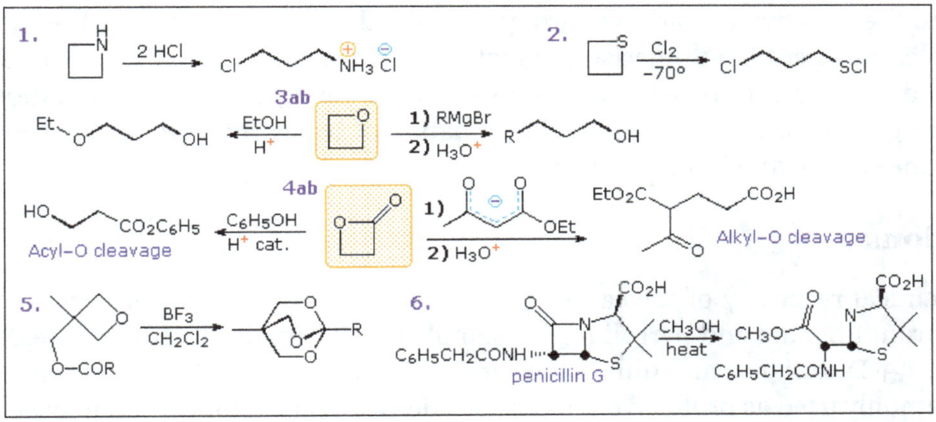

Five-Membered Rings

Preparation

Commercial preparation of furan proceeds by way of the aldehyde, furfural, which in turn is generated from pentose containing raw materials like corncobs, as shown in the uppermost equation below. Similar preparations of pyrrole and thiophene are depicted in the second row equations. Equation 1 in the third row illustrates a general preparation of substituted furans, pyrroles and thiophenes from 1,4-dicarbonyl compounds, known as the Paal-Knorr synthesis. Many other procedures leading to substituted heterocycles of this kind have been devised. Two of these are shown in reactions 2 and 3. Furan is reduced to tetrahydrofuran by palladium-catalyzed hydrogenation. This cyclic ether is not only a valuable solvent, but it is readily converted to 1,4-dihalobutanes or 4-haloalkylsulfonates, which may be used to prepare pyrrolidine and thiolane. Dipolar cycloaddition reactions often lead to more complex five-membered heterocycles.

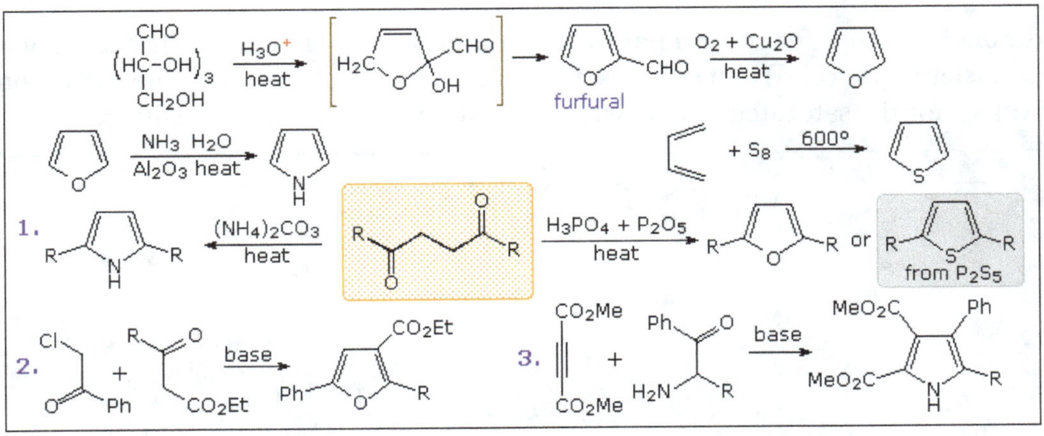

Indole is probably the most important fused ring heterocycle in this class. The first proceeds by an electrophilic substitution of a nitrogen-activated benzene ring. The second presumably takes place by formation of a dianionic species in which the $ArCH_2(-)$ unit bonds to the deactivated carbonyl group. Finally, the Fischer indole synthesis is a remarkable sequence of tautomerism, sigmatropic rearrangement, nucleophilic addition, and elimination reactions occurring subsequent to phenylhydrazone formation. This interesting transformation involves the oxidation of two carbon atoms and the reduction of one carbon and both nitrogen atoms.

Reactions

The chemical reactivity of the saturated members of this class of heterocycles: tetrahydrofuran, thiolane and pyrrolidine, resemble that of acyclic ethers, sulfides, and 2º amines. 1,3-Dioxolanes and dithiolanes are cyclic acetals and thioacetals. These units are commonly used as protective groups for aldehydes and ketones, and may be hydrolyzed by the action of aqueous acid.

It is the "aromatic" unsaturated compounds, furan, thiophene and pyrrole that require our attention. In each case the heteroatom has at least one pair of non-bonding electrons that may combine with the four π-electrons of the double bonds to produce an annulene having an aromatic sextet of electrons. This is illustrated by the resonance description at the top of the following diagram. The heteroatom Y becomes sp^2-hybridized and acquires a positive charge as its electron pair is delocalized around the ring. An easily observed consequence of this delocalization is a change in dipole moment compared with the analogous saturated heterocycles, which all have strong dipoles with the heteroatom at the negative end. As expected, the aromatic heterocycles have much smaller dipole moments, or in the case of pyrrole a large dipole in the opposite direction. An important characteristic of aromaticity is enhanced thermodynamic stability, and this is usually demonstrated by relative heats of hydrogenation or heats of combustion measurements. By this standard, the three aromatic heterocycles under examination are stabilized, but to a lesser degree than benzene.

Additional evidence for the aromatic character of pyrrole is found in its exceptionally weak basicity (pK_a ca. 0) and strong acidity (pK_a = 15) for a 2º amine. The corresponding values for the saturated amine pyrrolidine are: basicity 11.2 and acidity 32.

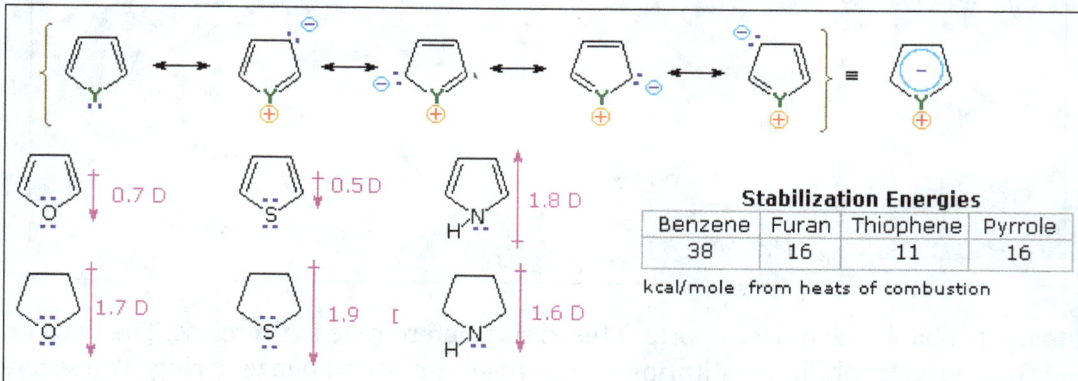

Stabilization Energies

Benzene	Furan	Thiophene	Pyrrole
38	16	11	16

kcal/mole from heats of combustion

Another characteristic of aromatic systems, of particular importance to chemists, is their pattern of reactivity with electrophilic reagents. Whereas simple cycloalkenes generally give addition reactions, aromatic compounds tend to react by substitution. As noted for benzene and its derivatives, these substitutions take place by an initial electrophile addition, followed by a proton loss from the "onium" intermediate to regenerate the aromatic ring. The aromatic five-membered heterocycles all undergo electrophilic substitution, with a general reactivity order: pyrrole >> furan > thiophene > benzene. Some examples are given in the following diagram. The reaction conditions show clearly the greater reactivity of furan compared with thiophene. All these aromatic heterocycles react vigorously with chlorine and bromine, often forming polyhalogenated products together with polymers. The exceptional reactivity of pyrrole is evidenced by its reaction with iodine (bottom left equation), and formation of 2-acetylpyrrole by simply warming it with acetic anhydride (no catalyst).

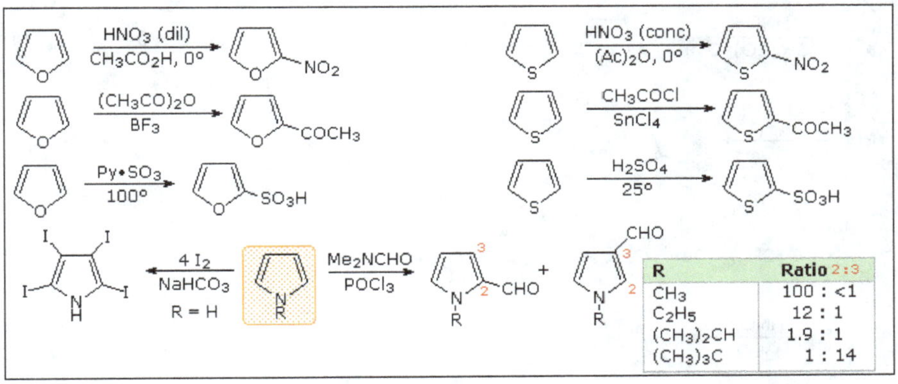

There is a clear preference for substitution at the 2-position (α) of the ring, especially for furan and thiophene. Reactions of pyrrole require careful evaluation, since N-protonation destroys its aromatic character. Indeed, N-substitution of this 2° amine is often carried out prior to subsequent reactions. For example, pyrrole reacts with acetic anhydride or acetyl chloride and triethyl amine to give N-acetylpyrrole. Consequently, the regioselectivity of pyrrole substitution is variable, as noted by the bottom right equation.

An explanation for the general α-selectivity of these substitution reactions is apparent from the mechanism outlined below. The intermediate formed by electrophile attack at C-2 is stabilized by charge delocalization to a greater degree than the intermediate from C-3 attack. The activation energy for substitution at the former position is less than the latter substitution.

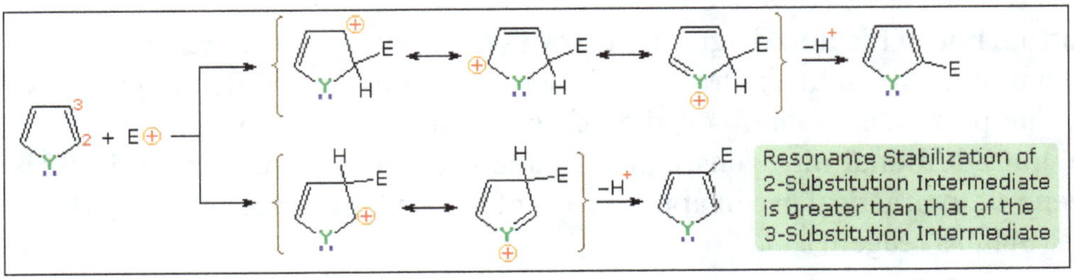

Functional substituents influence the substitution reactions of these heterocycles in much the same fashion as they do for benzene. Indeed, once one understands the ortho-para and meta-directing character of these substituents, their directing influence on heterocyclic ring substitution is not difficult to predict. The following diagram shows seven such reactions. Reactions 1 & 2 are 3-substituted thiophenes, the first by an electron donating substituent and the second by an electron withdrawing group. The third reaction has two substituents of different types in the 2 and 5-positions. Finally, figures' examples 4 through 7 illustrate reactions of 1,2- and 1,3-oxazole, thiazole and diazole. Note that the basicity of the sp^2-hybridized nitrogen in the diazoles is over a million times greater than that of the apparent sp^3-hybridized nitrogen, the electron pair of which is part of the aromatic electron sextet.

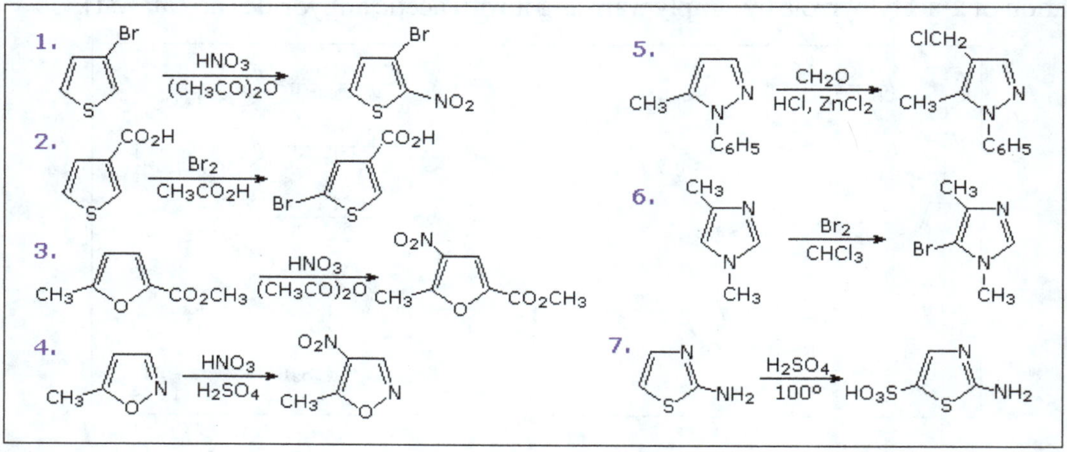

Other possible reactions are suggested by the structural features of these heterocycles. For example, furan could be considered an enol ether and pyrrole an enamine. Such functions are known to undergo acid-catalyzed hydrolysis to carbonyl compounds and alcohols or amines. Since these compounds are also heteroatom substituted dienes, we might anticipate Diels-Alder cycloaddition reactions with appropriate dienophiles. Furans may indeed be hydrolyzed to 1,4-dicarbonyl compounds, but pyrroles and thiophenes behave differently. The second two examples, shown in the middle, demonstrate typical reactions of furan and pyrrole with the strong dienophile maleic anhydride. The former participates in a cycloaddition reaction; however, the pyrrole simply undergoes electrophilic substitution at C-2. Thiophene does not easily react with this dienophile.

The bottom line of the new diagram illustrates the remarkable influence that additional nitrogen units have on the hydrolysis of a series of N-acetylazoles in water at 25 °C and pH=7. The pyrrole compound on the left is essentially unreactive, as expected for an amide, but additional nitrogens markedly increase the rate of hydrolysis. This effect has been put to practical use in applications of the acylation reagent 1,1'-carbonyldiimidazole (Staab's reagent).

Another facet of heterocyclic chemistry was disclosed in the course of investigations concerning the action of thiamine (following diagram). As its pyrophosphate derivative, thiamine is a coenzyme for several biochemical reactions, notably decarboxylations of pyruvic acid to acetaldehyde and acetoin. Early workers speculated that an "active aldehyde" or acyl carbanion species was an intermediate in these reactions. Many proposals were made, some involving the aminopyrimidine moiety, and others, ring-opened hydrolysis derivatives of the thiazole ring, but none were satisfactory. This puzzle was solved when R. Breslow (Columbia) found that the C-2 hydrogen of thiazolium salts was unexpectedly acidic (pK_a ca. 13), forming a relatively stable ylide conjugate base. As shown, this rationalizes the facile decarboxylation of thiazolium-2-carboxylic acids and deuterium exchange at C-2 in neutral heavy water.

Appropriate thiazolium salts catalyze the conversion of aldehydes to acyloins in much the same way that cyanide ion catalyzes the formation of benzoin from benzaldehyde, the benzoin condensation. Note that in both cases an acyl anion equivalent is formed and then adds to a carbonyl function in the expected manner. The benzoin condensation is limited to aromatic aldehydes, but the use of thiazolium catalysts has proven broadly effective for aliphatic and aromatic aldehydes. This approach to acyloins employs milder conditions than the reduction of esters to enediol intermediates by the action of metallic sodium.

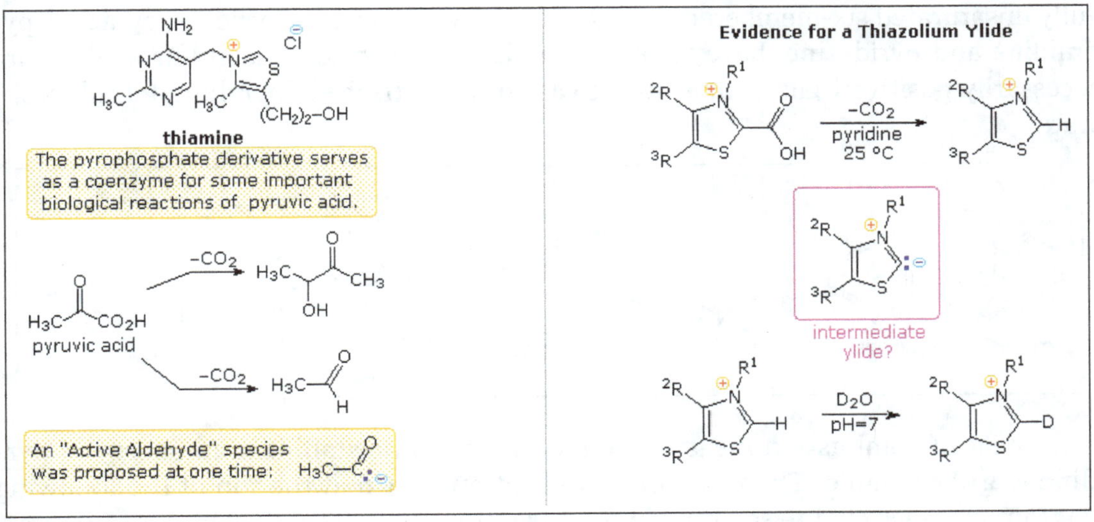

The most important condensed ring system related to these heterocycles is indole. Some electrophilic substitution reactions of indole are shown in the following diagram. Whether the indole nitrogen is substituted or not, the favored site of attack is C-3 of the heterocyclic ring. Bonding of the electrophile at that position permits stabilization of the onium-intermediate by the nitrogen without disruption of the benzene aromaticity.

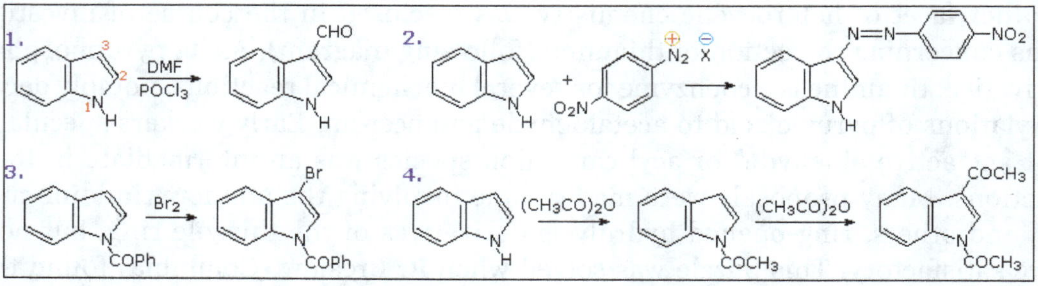

Six-Membered Rings

Properties

The chemical reactivity of the saturated members of this class of heterocycles: tetrahydropyran, thiane and piperidine, resemble that of acyclic ethers, sulfides, and 2° amines, and will not be described here. 1,3-Dioxanes and dithianes are cyclic acetals and thioacetals. These units are commonly used as protective groups for aldehydes and ketones, as well as synthetic intermediates, and may be hydrolyzed by the action of aqueous acid. The reactivity of partially unsaturated compounds depends on the relationship of the double bond and the heteroatom (e.g. 3,4-dihydro-2H-pyran is an enol ether).

Fully unsaturated six-membered nitrogen heterocycles, such as pyridine, pyrazine, pyrimidine and pyridazine, have stable aromatic rings. Oxygen and sulfur analogs are necessarily positively charged, as in the case of 2,4,6-triphenylpyrylium tetrafluoroborate.

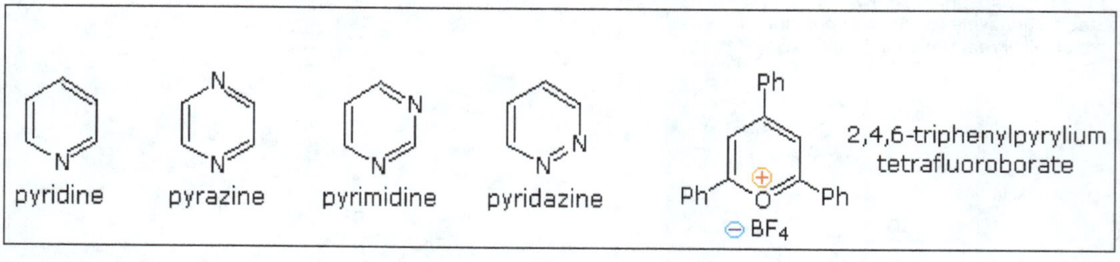

From heat of combustion measurements, the aromatic stabilization energy of pyridine is 21 kcal/mole. The resonance description drawn at the top of the following diagram includes charge separated structures not normally considered for benzene. The greater electronegativity of nitrogen (relative to carbon) suggests that such canonical forms may contribute to a significant degree. Indeed, the larger dipole moment of pyridine compared with piperidine supports this view. Pyridine and its derivatives are weak bases, reflecting the sp^2 hybridization of the nitrogen. From the polar canonical forms shown here, it should be apparent that electron donating substituents will increase the basicity of a pyridine, and that substituents on the 2 and 4-positions will influence this basicity more than an equivalent 3-substituent. The

pK_a values given in the table illustrate a few of these substituent effects. Methyl substituted derivatives have the common names picoline (methyl pyridines), lutidine (dimethyl pyridines) and collidine (trimethyl pyridines). The influence of 2-substituents is complex, consisting of steric hindrance and electrostatic components. 4-Dimethylaminopyridine is a useful catalyst for acylation reactions carried out in pyridine as a solvent. At first glance, the sp^3 hybridized nitrogen might appear to be the stronger base, but it should be remembered that N,N-dimethylaniline has a pK_a slightly lower than that of pyridine itself. Consequently, the sp^2 ring nitrogen is the site at which protonation occurs.

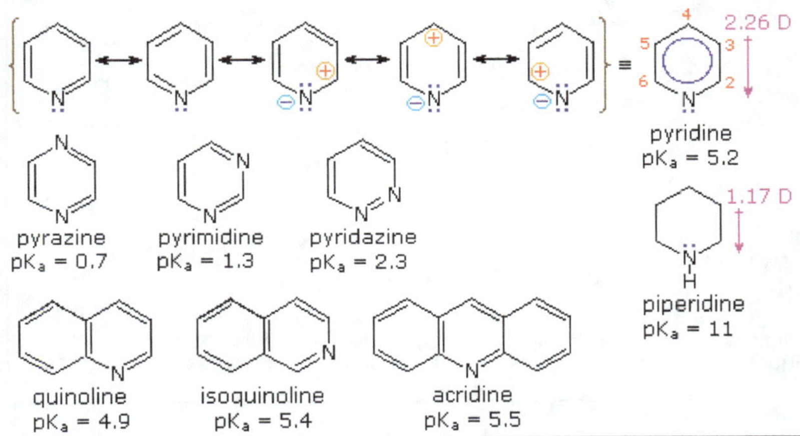

Pyridine Basicities

Substituent(s)	pK_a
4-methyl	6.0
3-methyl	5.7
2,6-dimethyl	6.7
2,4,6-trimethyl	7.5
2,6-di-*t*-butyl	3.6
3-methoxy	4.9
4-methoxy	6.5
4-dimethylamino	9.7
2-chloro	0.7
4-chloro	3.8
2,6-dichloro	-2.9

The diazines pyrazine, pyrimidine and pyridazine are all weaker bases than pyridine due to the inductive effect of the second nitrogen. However, the order of base strength is unexpected. A consideration of the polar contributors helps to explain the difference between pyrazine and pyrimidine, but the basicity of pyridazine seems anomalous. It has been suggested that electron pair repulsion involving the vicinal nitrogens destabilizes the neutral base relative to its conjugate acid.

Electrophilic Substitution of Pyridine

Pyridine is a modest base (pK_a=5.2). Since the basic unshared electron pair is not part of the aromatic sextet, as in pyrrole, pyridinium species produced by N-substitution retain the aromaticity of pyridine. As shown below, N-alkylation and N-acylation products may be prepared as stable crystalline solids in the absence of water or other reactive nucleophiles. The N-acyl salts may serve as acyl transfer agents for the preparation of esters and amides. Because of the stability of the pyridinium cation, it has been used as a moderating component in complexes with a number of reactive inorganic compounds. Several examples of these stable and easily handled reagents are shown at the bottom of the diagram. The poly (hydrogen fluoride) salt is a convenient source of HF for addition to alkenes and conversion of alcohols to alkyl fluorides, pyridinium chlorochromate (PCC) and its related dichromate analog are versatile oxidation agents and the tribromide salt is a convenient source of bromine. Similarly, the reactive compounds sulfur trioxide and diborane are conveniently and safely handled as pyridine complexes.

Amine oxide derivatives of 3° amines and pyridine are readily prepared by oxidation with peracids or peroxides, as shown by the upper right equation. Reduction back to the amine can usually be achieved by treatment with zinc (or other reactive metals) in dilute acid.

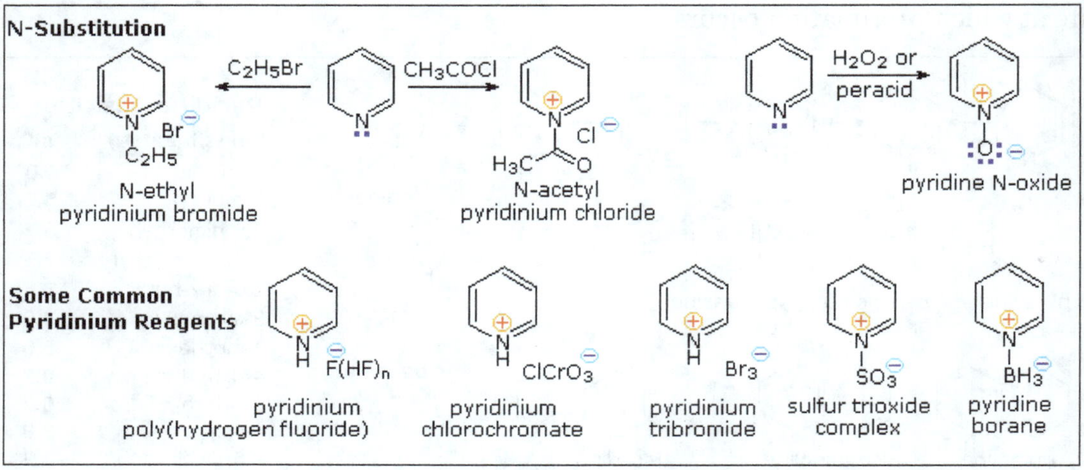

The electrophilic reagents and catalysts employed in these reactions coordinate with the nitrogen electron pair, exacerbating the positive charge at positions 2,4 & 6 of the pyridine ring. Three examples of the extreme conditions required for electrophilic substitution are shown on the left. Substituents that block electrophile coordination with nitrogen or reduce the basicity of the nitrogen facilitate substitution, as demonstrated by the examples in the blue-shaded box at the lower right, but substitution at C-3 remains dominant. Activating substituents at other locations also influence the ease and regioselectivity of substitution. The amine substituent in the upper case directs the substitution to C-2, but the weaker electron donating methyl substituent in the middle example cannot overcome the tendency for 3-substitution. Hydroxyl substituents at C-2 and C-4 tautomerize to pyridones, as shown for the 2-isomer at the bottom left.

Pyridine N-oxide undergoes some electrophilic substitutions at C-4 and others at C-3. The coordinate covalent N–O bond may exert a push-pull influence, as illustrated by the two examples on the right. Although the positively charged nitrogen alone would have a strong deactivating influence, the negatively charged oxygen can introduce electron density at C-2, C-4 & C-6 by π-bonding to the ring nitrogen. This is a controlling factor in the relatively facile nitration at C-4. However, if the oxygen is bonded to an electrophile such as SO_3, the resulting pyridinium ion will react sluggishly and preferentially at C-3.

The fused ring heterocycles quinoline and isoquinoline provide additional evidence for the stability of the pyridine ring. Vigorous permanganate oxidation of quinoline results in predominant attack on the benzene ring; isoquinoline yields products from cleavage of both rings. Note that naphthalene is oxidized to phthalic acid in a similar manner. By contrast, the heterocyclic ring in both compounds undergoes preferential catalytic hydrogenation to yield tetrahydroproducts. Electrophilic nitration, halogenation and sulfonation generally take place at C-5 and C-8 of the benzene ring, in agreement with the preceding description of similar pyridine reactions and the kinetically favored substitution of naphthalene at C-1 (α) rather than C-2 (β).

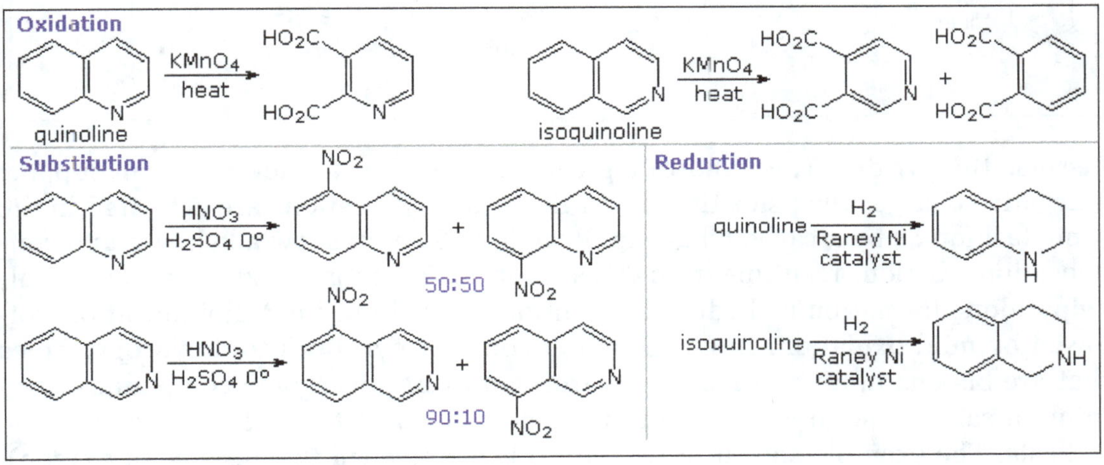

Other Reactions of Pyridine

Thanks to the nitrogen in the ring, pyridine compounds undergo nucleophilic substitution reactions more easily than equivalent benzene derivatives. In the following diagram, reaction 1 illustrates displacement of a 2-chloro substituent by ethoxide anion. The addition-elimination mechanism shown for this reaction is helped by nitrogen's ability to support a negative charge.

A similar intermediate may be written for substitution of a 4-halopyridine, but substitution at the 3-position is prohibited by the the failure to create an intermediate of this kind. The two Chichibabin aminations in reactions 2 and 3 are remarkable in that the leaving anion is hydride (or an equivalent). Hydrogen is often evolved in the course of these reactions. In accord with this mechanism, quinoline is aminated at both C-2 and C-4.

Addition of strong nucleophiles to N-oxide derivatives of pyridine proceed more rapidly than to pyridine itself, as demonstrated by reactions 4 and 5. The dihydro-pyridine intermediate easily loses water or its equivalent by elimination of the –OM substituent on nitrogen.

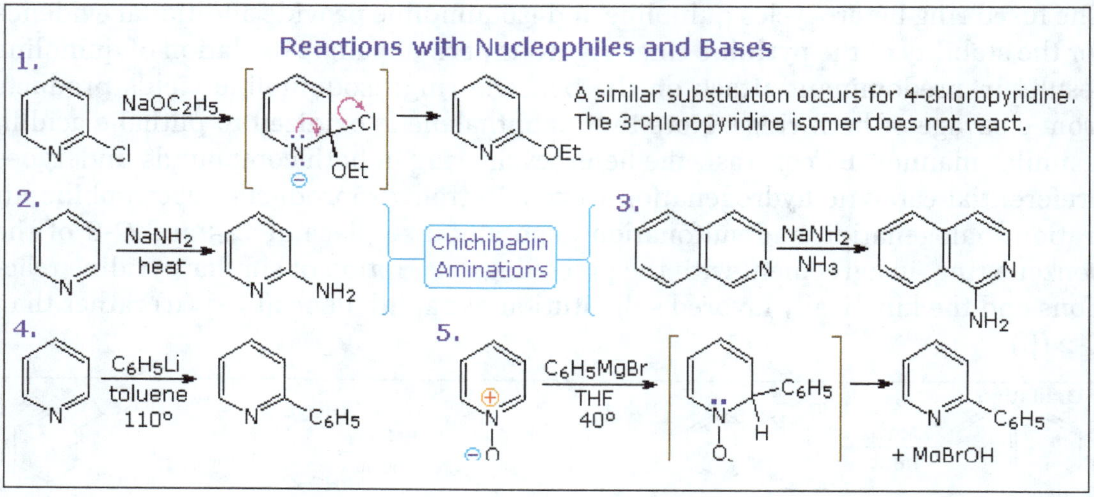

Because the pyridine ring (and to a greater degree the N-oxide ring) can support a negative charge, alkyl substituents in the 2- and 4-locations are activated in the same fashion as by a carbonyl group. Reactions 6 and 7 show alkylation and condensation reactions resulting from this activation. Reaction 8 is an example of N-alkylpyridone formation by hydroxide addition to an N-alkyl pyridinium cation, followed by mild oxidation. Birch reduction converts pyridines to dihydropyridines that are bis-enamines and may be hydrolyzed to 1,5-dicarbonyl compounds. Pyridinium salts undergo a one electron transfer to generate remarkably stable free radicals. The example shown in reaction 9 is a stable (in the absence of oxygen), distillable green liquid. Although 3-halopyridines do not undergo addition-elimination substitution reactions as do their 2- and 4-isomers, the strong base sodium amide effects amination by way of a pyridyne intermediate. This is illustrated by reaction 10. It is interesting that 3-pyridyne is formed in preference to 2-pyridyne. The latter is formed if C-4 is occupied by an alkyl substituent. The pyridyne intermediate is similar to benzyne.

Some Polycyclic Heterocycles

Heterocyclic structures are found in many natural products. Examples of some nitrogen compounds, known as alkaloids because of their basic properties. Some other examples are displayed in the following diagram. Camptothecin is a quinoline alkaloid which inhibits the DNA enzyme topoisomerase I. Reserpine is an indole alkaloid, which has been used for the control of high blood pressure and the treatment of psychotic behavior. Ajmaline and strychnine are also indole alkaloids, the former being an antiarrhythmic agent and latter an extremely toxic pesticide. The neurotoxins saxitoxin and tetrodotoxin both have marine origins and are characterized by guanidiniun moieties. Aflatoxin B1 is a non-nitrogenous carcinogenic compound produced by the Aspergillus fungus.

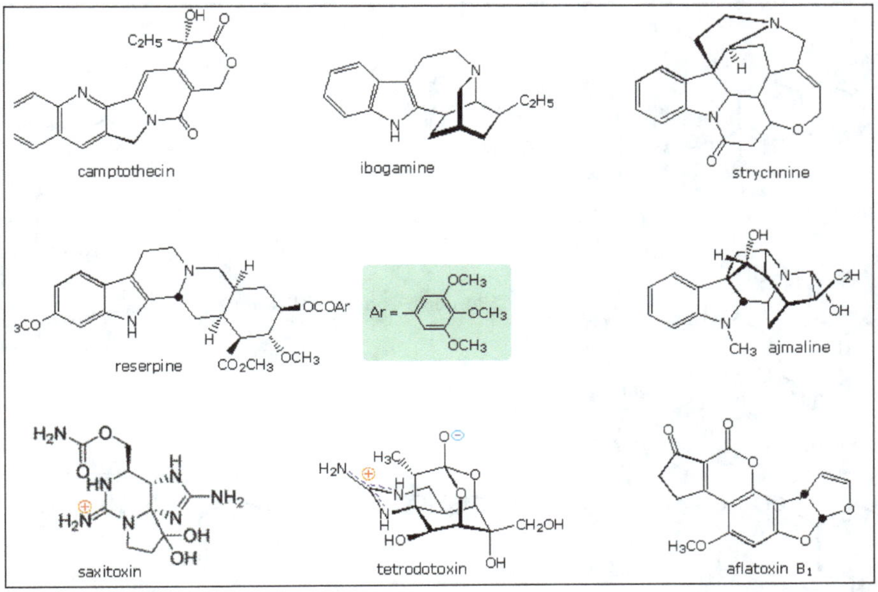

Porphyrin is an important cyclic tertrapyrrole that is the core structure of heme and chlorophyll. Derivatives of the simple fused ring heterocycle purine constitute an especially important and abundant family of natural products. The amino compounds adenine and guanine are two of the complementary bases that are essential components of DNA. Structures for these compounds are shown in the following diagram. Xanthine and uric acid are products of the metabolic oxidation of purines. Uric acid is normally excreted in the urine; an excess serum accumulation of uric acid may lead to an arthritic condition known as gout.

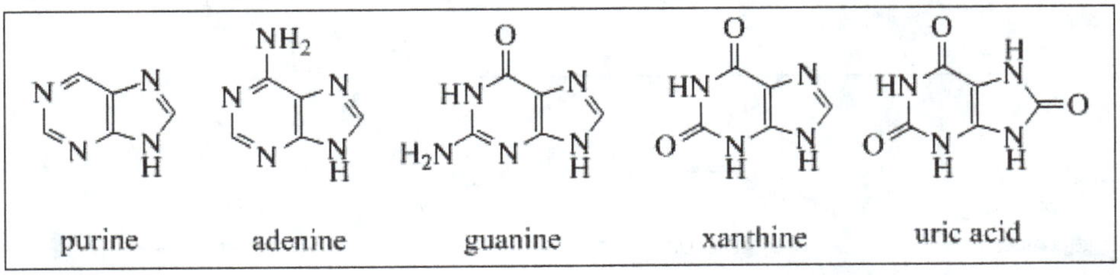

Caffeine, the best known of these, is a bitter, crystalline alkaloid. It is found in varying quantities, along with additional alkaloids such as the cardiac stimulants theophylline and theobromine in the beans, leaves, and fruit of certain plants. Drinks containing caffeine, such as coffee, tea and some soft drinks are arguably the world's most widely consumed beverages. Caffeine is a central nervous system stimulant, serving to ward off drowsiness and restore alertness. Paraxantheine is the chief metabolite of caffeine in the body.

Sulfur heterocycles are found in nature, but to a lesser degree than their nitrogen and oxygen analogs. Two members of the B-vitamin complex, biotin and thiamine, incorporate such heterocyclic moieties. These are shown together with other heterocyclic B-vitamins in the following diagram.

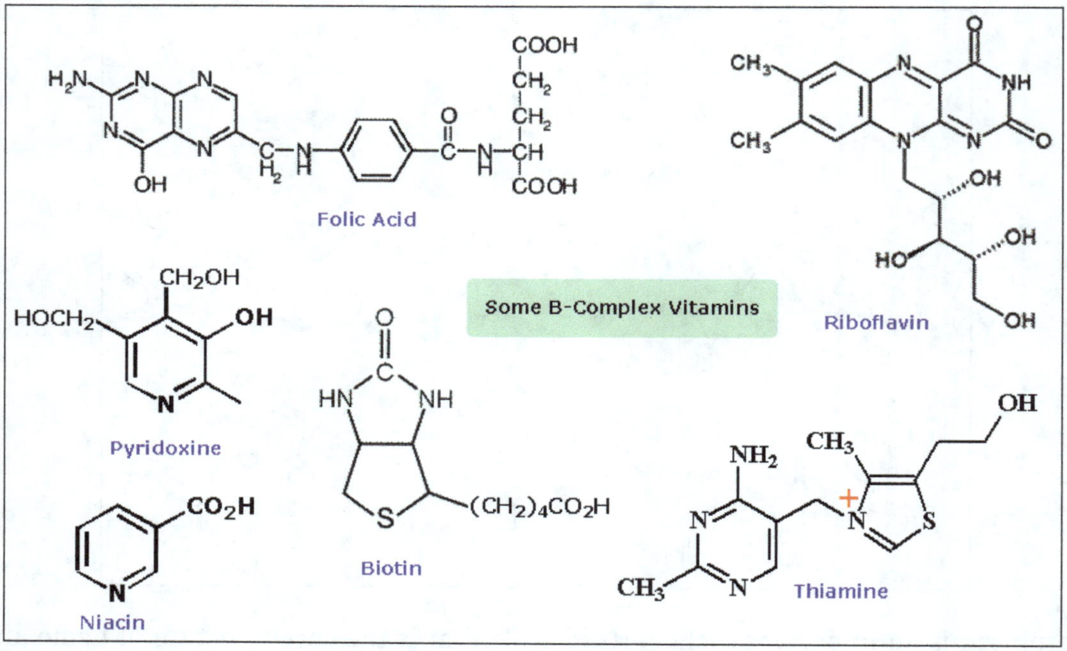

Terthienyl is an interesting thiophene trimer found in the roots of marigolds, where it provides nemicidal activity. Studies have shown that UV irradiation of terthienyl produces a general phototoxicity for many organisms. Polymers incorporating thiophene units and fused systems such as dithienothiophene have interesting electromagnetic properties, and show promise as organic metal-like conductors and photovoltaic materials. The charge transfer complex formed by tetrathiofulvalene and tetracyanoquinodimethane has one of the highest electrical conductivities reported for an organic solid.

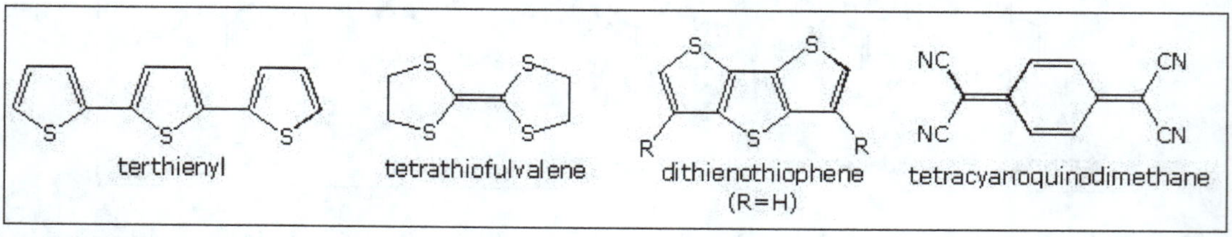

Polymer

Organic polymers are macromolecules composed of many repeating monomer units. Both synthetic and natural polymers play a crucial role in everyday life. Polysaccharides, polypeptides, and polynucleotides are the main types of biopolymers in living cells. These polymers are synthesized by enzyme-mediated processes in cells.

Organic Polymer Carrier

Organic polymer carriers include natural polymer carriers and synthetic organic polymer carriers.

- Natural polymer carriers are generally nontoxic to microorganisms, have good mass transfer performance, but low strength and can easily be broken down by microorganisms in anaerobic conditions. Common natural polymer carriers are agar, gelatin, carrageenan, sponges, chitin, alginate, and chitosan. The agar-embeding cell method is simple and nontoxic to cells, with a larger clearance, allowing the diffusion of high molecular substances, but the mechanical strength and chemical stability is not good.

  Alginate is a widely used immobilization carrier. It has good chemical stability, nontoxic, high efficiency, and is suitable for fixing living cells or sensitive cells, but it cannot resist phosphate and Na^+, K^+, Mg^{2+}, and other cations with high concentrations, and is broken easily and dissolved. The gel strength is not enough and it is hard to repeat the use of immobilized cells.

- Synthetic organic polymer gel carriers show good resistance to antimicrobial decomposition, high mechanical strength, and good chemical stability. However, the mass transfer performance is poor and the cell activity will reduce in the embedding process. Common synthetic organic polymer gel carriers are polyacrylamides, light-curing resins, polyvinyl alcohol, and polyacrylic acid gel.

Natural Organic Polymer

The polymers which are obtained naturally are called natural polymers. A natural polymer has its origin in plants and animals. Starch, cellulose, proteins, natural rubber etc. are the examples of natural polymers.

Classification of Natural Polymers

- Plant origin: Cellulose, Hemicellulose, Glucomannan, Agar, Starch, Pectin, Inulin, Rosin, Guar gum, Locust bean Gum, Gum Acacia, Karaya gum, Gum Tragacanth, Aloe Vera gel.
- Animal origin: Chitin, Alginates, Carageenans, Psyllium, Xanthum gum.

Natural Polymers from Plant Origin

Cellulose

Cellulose was discovered in 1838 by the French chemist Anselme Payen, who isolated it from plant matter and determined its chemical formula. Cellulose is an organic polysaccharide with the formula $(C_6H_{10}O_5)_n$, consisting of a linear chain of several hundred to over ten thousand $\beta(1\rightarrow4)$ linked D-glucose units.

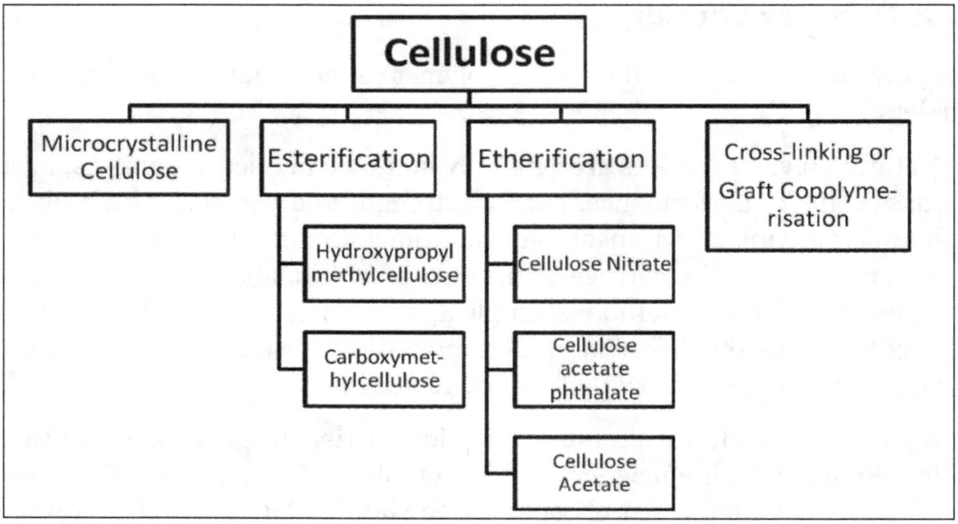

Cellulose derivatives.

The polysaccharides of the plant cell wall consist mainly of cellulose, hemicelluloses and pectin. Cellulose is an essential structural component of cell walls in higher plants and is the most abundant organic polymer on earth. Many parallel cellulose molecules form crystalline microfibrils that are mechanically strong and highly resistant to enzymatic attack. These are aligned with each other to provide structure to the cell wall. Cellulose is insoluble in water and indigestible by the human body.

Microcrystalline cellulose is mainly used in the pharmaceutical industry as a diluent/binder in tablets for both the granulation and direct compression processes. Controlled release applications for cellulose derivatives include the formulation of membrane controlled drug release systems or monolithic matrix systems. Film coating techniques for the manufacture of membrane controlled release systems include enteric coated dosage forms and the use of semipermeable membranes in osmotic pump delivery systems.

Hydroxypropylmethyl cellulose is a partly Omethylated and *O*-(2-hydroxypropylated) cellulose ether derivative that has been extensively investigated as an excipient in controlled release drug delivery systems due to its gel forming ability. In a study where two cellulose ethers; hydroxypropylmethylcellulose and carboxymethylcellulose were employed as polymeric carrier materials in matrix tablets for controlling the release of a soluble drug, diltiazem, it was found that each polymer on its own could sustain drug release over an extended period of time in these systems.

More importantly, a mixture of the two cellulose ethers in the matrix type tablets enabled zero order drug release kinetics at both pH 4.5 and 6.8. Hydroxypropylmethyl cellulose monolithic matrix systems showed similar dissolution profiles as a commercial osmotic pump system for glipizide, a drug with low solubility. It was further found that the hydroxypropylmethyl cellulose matrix systems have a stronger gel structure than those made of polyethylene oxide, which may provide superior in vivo performance in terms of matrix resistance to the destructive forces within the gastro-intestinal tract.

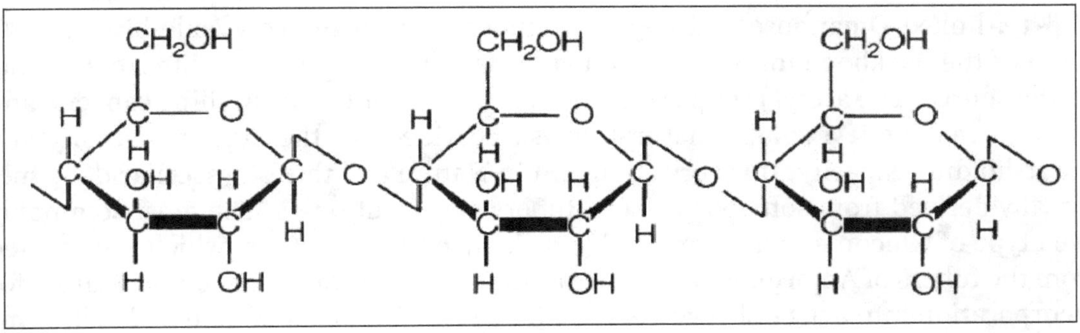

Structure of Cellulose.

Hemicellulose

A hemicellulose is a heteropolymer (matrix polysaccharides), such as arabinoxylans, present along with cellulose in almost all plant cell walls. While cellulose is crystalline, strong, and resistant to hydrolysis, hemicellulose has a random, amorphous structure with little strength.

Unlike cellulose, hemicellulose (also a polysaccharide) consists of shorter chains - 500-3,000 sugar units. In addition, hemicellulose is a branched polymer, while cellulose is unbranched. Hemicellulose polysaccharides consist of xyloglucans, xylans and mannans that can be extracted from the plant cell wall with a strong alkali. They have backbones made up of β-1,4-linked Dglycans. Xyloglucan has a similar backbone as cellulose, but contains xylose branches on 3 out of every 4 glucose monomers. The β-1,4-linked DXylan backbone of arabinoxylan contains arabinose branches.

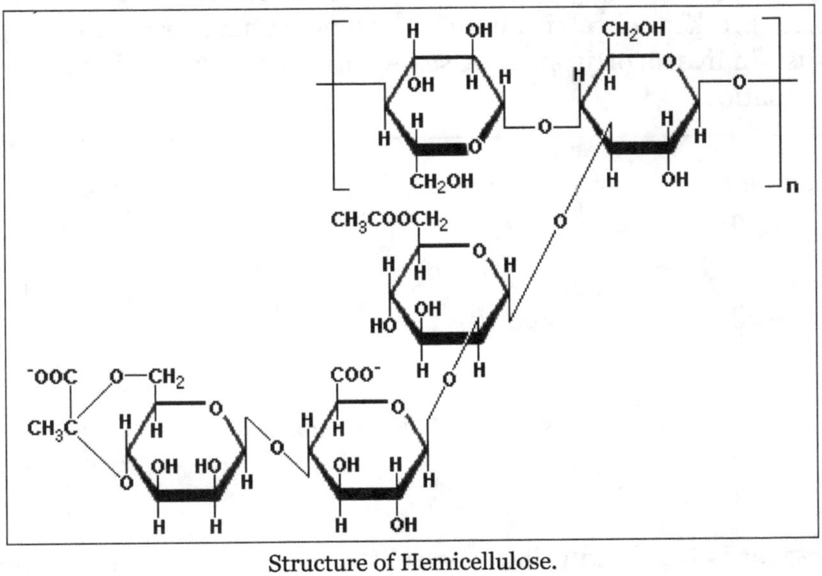

Structure of Hemicellulose.

Glucomannan

Glucomannan is a hydrocolloidal polysaccharide of the mannan family consisting of β-1,4 linked Dmannose and D-glucose monomers (with acetyl side branches on some of the backbone units), but the mannose: glucose ratio may differ depending on the source. The acetyl groups contribute to its solubility and swelling capacity and assist in making it a soluble natural polysaccharide with the highest viscosity and water-holding capacity. It is very abundant in Nature and this polysaccharide is specifically derived from softwoods, roots, tubers and plant bulbs. The most commonly used type of Glucomannan is referred to as konjac Glucomannan, which is extracted from the tubers of Amorphophallus konjac and is a very promising polysaccharide for incorporation into drug delivery systems. Since konjac Glucomannan by itself forms very weak gels, it has been investigated as an effective exponent in controlled release drug delivery devices in combination with other polymers or by modifying its chemical structure.

It was shown that konjac Glucomannan gel systems were able to maintain the integrity and control the release of theophylline and diltiazem for 8 hours. This was, however, dependent on the country of origin (i.e. Japan, Europe or America) due to differences in the degree of acetylation of the konjac Glucomannan.18 Matrix tablets prepared from konjac glucomannan alone showed the ability to sustain the release of cimetidine in the physiological environments of the stomach and small intestines but the presence of β-mannanase (colon) accelerated the drug release substantially. Mixtures of konjac Glucomannan and xanthan gum in matrix type tablets showed high potential to sustain and control the release of the drug due to stabilization of the gel phase of the tablets by a network of intermolecular hydrogen bonds between the two polymers to effectively retard drug diffusion. Konjac Glucomannan was used to form hydrophilic cylinders and particles for controlled release of DNA. Konjac Glucomannan cross-linked with trisodium trimetaphosphate formed hydrogel systems that could sustain hydrocortisone release dependent on crosslinking density and enzymatic degradation.

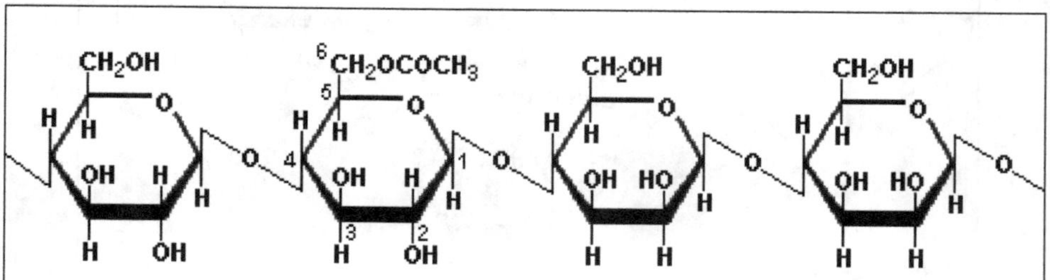

Structure of Glucomannan.

Agar

Agar or agar-agar is the dried gelatinous substance obtained from Gelidium amansii

(Gelidaceae) and several other species of red algae like, grailaria (Gracilariaceae) and Pterocladia (Gelidaceae). Agar consists of a mixture of agarose and agaropectin. The predominant component, agarose, is a linear polymer, made up of the repeating monomeric unit of agarobiose. Agarobiose is a disaccharide made up of D-galactose and 3,6- anhydro-L-galactopyranose.

Agaropectin is a heterogeneous mixture of smaller acidic molecules that gel poorly. Its great gelling power in an aqueous environment allows it to form gels which are more resistant (stronger) than those of any other gelforming agent, assuming the use of equal concentrations. It can be used over a wide range of pH, from 5 to 8, and in some cases beyond these limits. It withstands thermal treatments very well, even above 100°C which allows good sterilization. A 1.5% aqueous solution gels between 32 °C-43 °C and does not melt below 85 °C. This is a unique property of agar, compared to other gelling agents. Agar gives gels without flavour and does not need the additions of cations with strong flavours (potassium or calcium) it can be used without problems to gel food products with soft flavours. Its gel has an excellent reversibility allowing it to be repeatedly gelled and melted without losing any of the original properties. Agar is used as Suspending agent, emulsifying agent, gelling agent in suppositories, surgical lubricant, tablet disintegrants, medium for bacterial culture, laxative.

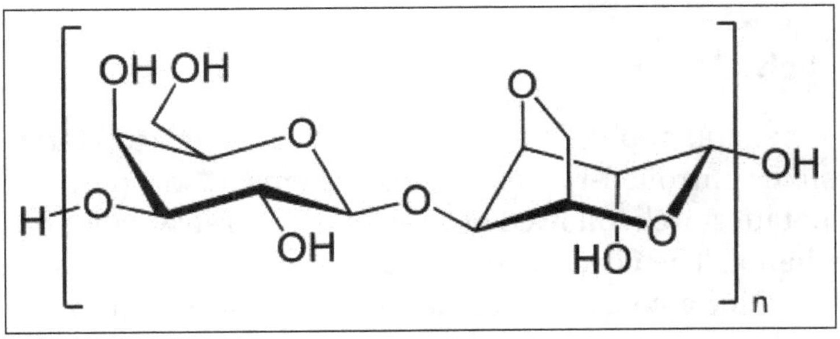

Structure of Agar.

Starch

Starch or amylum is a carbohydrate consisting of a large number of glucose units joined together by glycosidic bonds. This polysaccharide is produced by all green plants as an energy store. It is the principal form of carbohydrate reserve in green plants and especially present in seeds and underground organs. Starch occurs in the form of granules (starch grains). A number of starches are recognized for pharmaceutical use. These include maize (Zea mays), rice (Oryza sativa), wheat (Triticum aestivum), and potato (Solanum tuberosum).

It is comprised of two polymers, namely amylose (a non-branching helical polymer consisting of α-1, 4 linked D-glucose monomers) and amylopectin (a highly branched polymer consisting of both α-1,4 and α-1,6 linked D-glucose monomers).

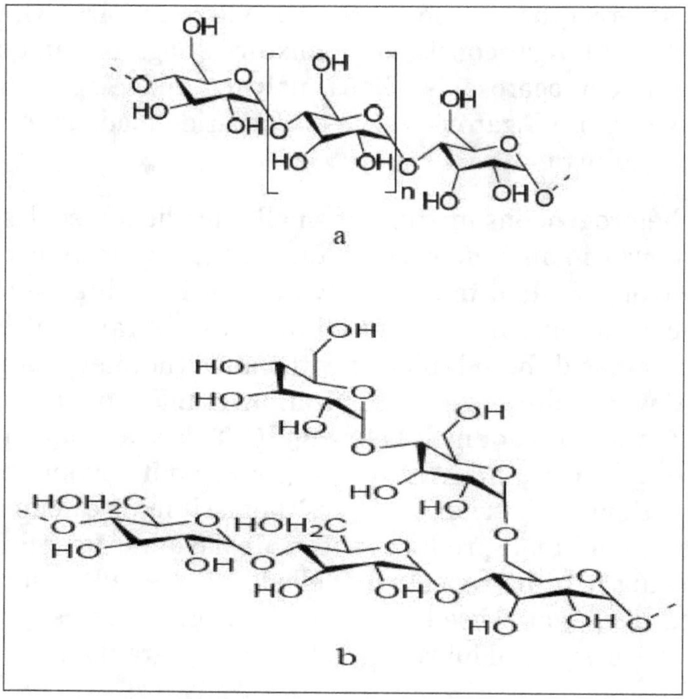

Structure of a) Amylose b) Amylopectin.

Modified Starch

It was tested for general applicability of a new pre-gelatinized starch product in directly compressible controlled-release matrix systems. It was prepared by enzymatic degradation of potato starch followed by precipitation (retrogradation), filtration and washing with ethanol. The advantages of the material include ease of tablet preparation, the potential of a constant release rate (zero-order) for an extended period of time and its ability to incorporate high percentages of drugs with different physicochemical properties. Release rates from retrograded pre-gelatinized starch tablets can be enhanced or decreased to the desired profile by different parameters like geometries of the tablet, compaction force and the incorporation of additional excipients.

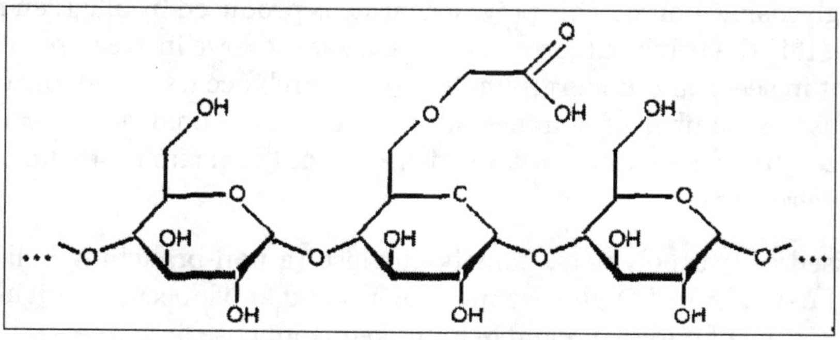

Structure of Modified Starch.

Native Starch

It may not be suitable in controlled release drug delivery systems due to substantial swelling and rapid enzymatic degradation resulting in too fast release of many drugs. This has led to the use of derivatives of starch that are more resistant to enzymatic degradation as well as crosslinking and formation of co-polymers. Starch acetate prepared by acetyl esterification has shown retarded enzymatic degradation with the potential to be used as a colon-targeted drug delivery carrier. High amylase carboxymethyl starch produced by spray drying showed a high loading capacity for the soluble drug, acetaminophen, in controlled release direct compressible matrix systems.

To deliver proteins or peptide drugs orally, microcapsules containing a protein and a proteinase inhibitor were prepared. Starch/bovine serum albumin mixed-walled microcapsules were prepared using interfacial cross-linking with terephthaloyl chloride. The microcapsules were loaded with native or amino-protected aprotinin by incorporating protease inhibitors in the aqueous phase during the cross-linking process. The protective effect of microcapsules with aprotinin for bovine serum albumin was revealed in vitro.

Pectin

Pectin is the purified carbohydrate product obtained by acid hydrolysis from the inner portion of the rind of citrus peels i.e. Citrus Simon or Citrus Aurantium, (Rutaceae). The main component of pectin is a linear polysaccharide composed of α-1,4-linked D-galacturonic acid residues interrupted by 1,2- linked L-rhamnose residues with a few hundred to about one thousand building blocks per molecule, corresponding to an average molecular weight of about 50,000 to about 1,80,000. The galacturonic acid polysaccharides are rich in neutral sugars such as rhamnose, arabinose, galactose, xylose and glucose. The composition of pectin can vary based on the botanical source, for example pectin from citrus contains less neutral sugars and has a smaller molecular size as compared to pectin obtained from apples.

Pectin has been investigated as an excipient in many different types of dosage forms such as film coating of colon-specific drug delivery systems when mixed with ethyl cellulose, micro-particulate delivery systems for ophthalmic preparations and matrix type transdermal patches. It has high potential as a hydrophilic polymeric material for controlled release matrix drug delivery systems, but its aqueous solubility contributes to the premature and fast release of the drug from these matrices.

It was investigated that the suitability of amidated pectin as a matrix patch for transdermal chloroquine delivery in an effort to mask the bitter taste when orally administered. The results suggest that the pectin-chloroquine patch matrix preparation has potential applications for the transdermal delivery of chloroquine and perhaps in the management of malaria. Calcium pectinate nanoparticles to deliver insulin were prepared as a potential colonic delivery system by ionotropic gelation.

Micro particulate polymeric delivery systems have been suggested as a possible approach to improve the low bioavailability characteristics shown by standard ophthalmic vehicles (collyria). In this context pectin microspheres of piroxicam were prepared by the spray drying technique. In vivo tests in rabbits with dispersions of piroxicam-loaded microspheres also indicated a significant improvement of piroxicam bioavailability in the aqueous humour (2.5-fold) when compared with commercial piroxicam eye drops.

Depending on the type and structure of the pectin molecule, pectins can gel in various ways. Gelling can be induced by acid or cross-linking with calcium ion or by reaction with alginate. When a pectin solution is titrated with acid, the ionization of carboxylate groups on pectins is repressed causing pectin molecules to no longer repel each other over their entire chains. The pectins can thus associate over a portion of their chains to form acid-pectin gels. Gel forming systems have been investigated widely for sustained drug delivery. A mixture of xyloglucan with pectin resulted in an in situ gel forming system with sustained paracetamol drug delivery in rats.

In relation to cosmetics, using citronella as a model compound, pectin gel formulations were evaluated for controlled fragrance release by kinetic and static methods. These formulations showed a prolonged duration of fragrance release and limitation of fragrance adsorption to the receptor skin layers. The increase in pectin concentrations suppressed the fragrance release by a diffusion mechanism, thereby proving that pectin/calcium microparticles are promising materials for controlled fragrance release.

In relation to the food industry, folic acid incorporated microcapsules were prepared using alginate and combinations of alginate and pectin polymers so as to improve stability of folic acid. Folic acid stability was evaluated with reference to encapsulation efficiency, gelling and hardening of capsules, capsular retention during drying and storage. The blended alginate and pectin polymer matrix increased the folic acid encapsulation efficiency and reduced leakage from the capsules as compared to those made with alginate alone, they showed higher folic acid retention after freeze drying and storage.

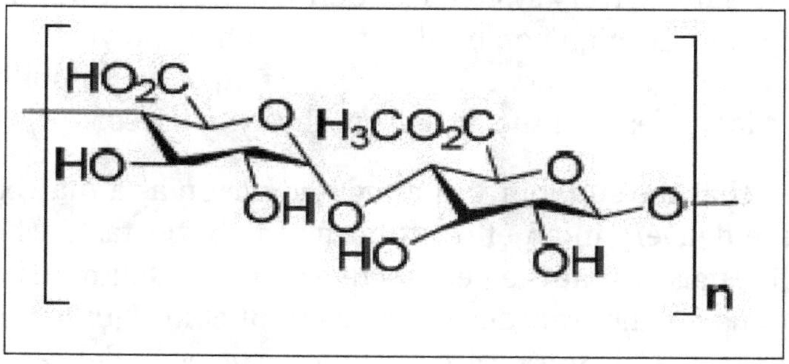

Structure of Pectin.

Inulin

It is a polysaccharide from the bulbs of Dehlia, Inula Helenium (Composite), roots of Dendelion, Taraxacum officinale (Composite). Burdock root, Saussurea lappa (Composite) or chicory roots, Cichonium intybus (Composite).

Inulin consists of a mixture of oligomers and polymers that belong to the group of gluco-fructans and occur in plants such as garlic, onion, artichoke and chicory. The inulin molecules contain from two to more than 60 fructose molecules linked by β-2,1- bonds. Inulin is resistant to digestion in the upper gastrointestinal tract, but is degraded by colonic microflora.

Inulin with a high degree of polymerization was used to prepare biodegradable colon-specific films in combination with Eudragit RS that could withstand break down by the gastric and intestinal fluids. It was shown in another study where different Eudragits were formulated into films with inulin that when a combination of Eudragit RS and Eudragit RL was mixed with inulin it exhibited better swelling and permeation properties in colonic medium rather than other gastrointestinal media. Methylated inulin hydrogels were developed as colon-specific drug delivery systems and investigated for water uptake and swelling. The hydrogels exhibited a relatively high rate of water uptake and anomalous dynamic swelling behaviour.

Inulin derivatised with methacrylic anhydride and succinic anhydride produced a pH sensitive hydrogel by UV irradiation that exhibited a reduced swelling and low chemical degradation in acidic medium, but it had a good swelling and degradation in simulated intestinal fluid in the presence of its specific enzyme, inulinase.

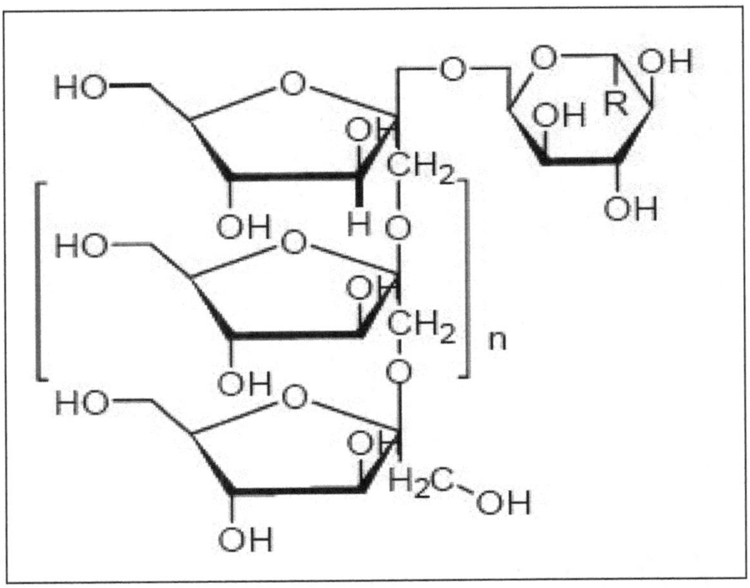

Structure of Inulin.

References

- Classification-organic-compounds, chemistry: byjus.com, Retrieved 11, July 2020
- Alicyclic-compound, science: britannica.com, Retrieved 09, January 2020
- Aromatic-compounds, chemistry: byjus.com, Retrieved 15, August 2020
- Heterocy, reusch: chemistry.msu.edu, Retrieved 06, April 2020
- Organic-polymer, chemical-engineering: sciencedirect.com, Retrieved 28, June 2020
- Natural-Polymers-A-comprehensive-Review-236217541: researchgate.net, Retrieved 18, March 2020

3

Organic Reactions and their Types

The chemical reactions involving organic compounds are organic reactions. There are many basic organic chemistry reactions like substitution reactions, addition reactions, elimination reactions, pericyclic reactions, rearrangement reactions and redox reactions. This chapter discusses in detail various types of reactions.

Organic reactions are chemical reactions involving organic compounds. The basic organic chemistry reaction types are addition reactions, elimination reactions, substitution reactions, pericyclic reactions, rearrangement reactions and redox reactions. In organic synthesis, organic reactions are used in the construction of new organic molecules. The production of many man-made chemicals such as drugs, plastics, food additives, fabrics depend on organic reactions.

Classification of Organic Reactions

Organic chemistry has a strong tradition of naming a specific reaction to its inventor or inventors and a long list of so-called named reactions exists, conservatively estimated at 1000. A very old named reaction is the Claisen rearrangement and a recent named reaction is the Bingel reaction. When the named reaction is difficult to pronounce or very long as in the Corey-House-Posner-Whitesides reaction it helps to use the abbreviation as in the CBS reduction. The number of reactions hinting at the actual process taking place is much smaller, for example the ene reaction or aldol reaction.

Another approach to organic reactions is by type of organic reagent, many of them inorganic, required in a specific transformation. The major types are oxidizing agents such as osmium tetroxide, reducing agents such as Lithium aluminium hydride, bases such as lithium diisopropylamide and acids such as sulfuric acid.

Fundamentals

Factors governing organic reactions are essentially the same as that of any chemical reaction. Factors specific to organic reactions are those that determine the stability of reactants and products such as conjugation, hyperconjugation and aromaticity and the presence and stability of reactive intermediates such as free radicals, carbocations and carbanions.

An organic compound may consist of many isomers. Selectivity in terms of regioselectivity, diastereoselectivity and enantioselectivity is therefore an important criterion for many organic reactions. The stereochemistry of pericyclic reactions is governed by the Woodward-Hoffmann rules and that of many elimination reactions by the Zaitsev's rule.

Substitution Reaction

Substitution reaction are also called displacement reactions.These are the reactions in which an atom or a group of atoms attached to a carbon atom in a molecule is replaced by some other atom or group of atoms without any change in the structure of the remaining part of the molecule. The product formed is known as the substitution product. The new atom or group which enters the molecule is called the substituent.

Some example of substitution reactions are:

$CH_4 + cl_2 \rightarrow CH_3cl + Hcl$

- Methane Methyl chloride: Here H-atom of CH_4 is replaced by chloride atom.

$CH_3 - CH_2 Br + KOH (aq) \rightarrow CH_3CH_2OH + KBr$

- Ethyl bromide Ethyl alcohol: Here Br atom of ethyl bromide is replaced by OH group.

$CH_3 -CH_2 = CH_2 \rightarrow ClCH_2 -CH=CH_2 + Hcl$

- Propene Allyl chloride: Here H-atom of CH_3 group is replaced by cl-atom.

Types of Substitution Reaction

Depending upon the nature of attacking reagent, substitution reactions can be classified as below:

Electrophilic Substitution Reactions

These are the reactions in which an atom or group in a molecule is replaced by an electrophile. Such reactions are shown by aromatic compound.

$CH_5 + HONO_2 \rightarrow CH_5NO_2 + H_2O$

$CH_5 + NO_2 \rightarrow CH_5 NO_2 + H_2O\ (H_2sO_4)$

Here Nitronium ion (NO_2 ions)acts as an electrophile and the process is called nitration. Mechanism: Consider the chlorination of benzene in presence of halogen carrier($Fecl_3$).

$Fecl_3 + cl\text{-}cl \rightarrow Fecl_4^- + cl$(chloronium ion)

Electrophiles are involved in electrophilic substitution reactions and particularly in electrophilic aromatic substitutions. Electrophilic reactions to other unsaturated compounds than arenes generally lead to electrophilic addition rather than substitution.

Necleophilic Substitution Reactions

These are the reaction sin which an atom or a group in a molecule is replaced by a nuclophile. Such reactions are shown by alkyl halides. Consider the action of aqueous KOH on methyl chloride.

$$HO^{-}+H\text{-}CH_3\ Cl\rightarrow HO\text{-}CH_3OH + Cl^{-}$$

A nucleophile (literally "nucleus lover") is a molecule or ion capable of acting as a Lewis base (i.e., an electron pair donor). Nucleophiles can be described as "electron-rich" while their targets or substrates can be described as "electron-deficient" (they are Lewis acids, i.e., electron pair acceptors). In a nucleophilic substitution the nucleophile takes the place of–or substitutes for–some atom or group on the substrate (called the "leaving group"):

$$Nu: + R:L \rightarrow R:Nu^{+} + :L^{-}$$

Nucleophile Substrate Product Leaving Group

If the nucleophile is neutral the product will be charged since the leaving group takes both bonding electrons away with it. If the nucleophile is an anion then the product will be neutral:

$$Nu: - + R:L \rightarrow R:Nu + :L^{-}$$

Stronger bases make better nucleophiles (e.g., OH- is a better nucleophile than H2O). "Good" substrates include cations, central atoms with incomplete octets or double bonds (like sp2 carbons) or carbons with partial positive charges. Halogens are generally more electronegative than carbon and so organohalogen compounds are usually subject to nucleophilic attack at the carbon attached to the halogen (which would be the positive end of a dipole). For example, 2-chloro-2-methylpropane (commonly known as t-butyl chloride) will undergo nucleophilic substitution with hydroxide ion:

$$(CH_3)3CCl + OH - \rightarrow (CH_3)3COH + Cl^{-}$$

This is a typical synthetic route for producing an alcohol from an alkyl halide.

Nucleophilic substitution reactions have been studied for many years. It was noticed fairly early that while the overall reaction was similar in the vast majority of cases the kinetics of the process was not always the same. In some substitutions the concentration of the nucleophile had no effect on the rate. In others, the rate was directly proportional to the concentration of the nucleophile. This suggested that two different mechanisms must be at work. The factor which determines the mechanism employed is typically the nature of the substrate itself and NOT the particular nucleophile. Neucleophilic substitution reactions are further classified as:

Neucleophilic Substitution Biomolecular

Such reactions are shown by primary alkyl halides and involve a single step. The breaking of C-X bond and the making of C-OH bond takes place simultaneously. The neucleophile approaches the C-atom from the side opposite to that carrying the halogen. In such reactions, an inverted product is formed. The reaction involves the formation of a transition state.

The reaction follows the rate law, Rate=k[Alkyl halide][OH]. As the slow rate determining step involves two molecule (Alkyl halide and alkali); therefore, the reaction is known as bimolecular substitution reaction.

Most necleophilic substitutions, which involve the expulsion of an originally neutral substituent, notably of halogen, from the aromatic ring, at temperature which are not particularly high, use the biomolecular mechanism SN2. This is established by their second order kinetics, which are documented by many records:

$$(NO_2)_2CH_3{\bullet}Cl+OEt^- \rightarrow (NO_2)_2CH_3{\bullet}OEt+Cl^-$$

In substitution of this type,the rate of attack by different reagent on the same aromatic molecules follow the general order of nucleophilic strength towards carbon. This is the conclusion to which Bunnatt and Zahler come after having assembled data from many sources. The first mechanism is known as SN1 (substitution, nucleophilic, unimolecular) because only one molecule is involved in the first step–the rate determining step. Reactions occurring by this mechanism should exhibit first-order kinetics, i.e., the rate law should have the form "rate = k[substrate]1". Because the nucleophile is not involved until after the slow step its concentration will have no effect on the rate.

The alternate mechanism is called SN2 (substitution, nucleophilic, bimolecular) because two molecules are involved in the rate determining (and only) step. Such reactions exhibit overall second-order kinetics. The rate is proportional to both the concentration of the substrate and the concentration of the nucleophile. Reactions like this will have a rate law in the form "rate = k[substrate] [nucleophile]".

Nucleophilic Substitution Unimolecular

This mechanism is generally followed by tertiary alkyl halide. In the first step, tertiary alkyl halie breaks hydrolytically to form intermediate carbonium ion. The formation of carbonium ion is the slow rate determine step.

In the second step, the nucleophile attacks the carbonium ion to form an alcohol. Consider the action of aqueous KOH or tertiary Butyl bromide.

CH_3 CH_3

CH_3 ____ C—-Br ⇄ CH_3 ——C +Br^- (Slow Step)

CH_3 CH_3

CH_3 CH_3

CH_3 ____ C + OH^- ⇄ CH_3 ——C +OH (Fast Step)

CH_3 CH_3

(carbonium ion) Tert. butyl alcohol

The best established example of nucleophilic aromatic substitution by the unimolecular mechanism, SN1, is the uncatalysed decomposition of diazoium ions, in hydroxylic solvent, to give phenols or phhenolic ethers, accomplished often byaryl halides or others such substitution products, if the necessary necleophilic anions are present in the solution:

- ArN_2+ → Ar + N_2 (Slow)
- Ar + H_2o →Ar• OH+H (fast)
- Ar + ROH→ Ar•OR + H (Fast)
- Ar +cl^-→ Ar• Cl (Fast)

These are SN1 mechanism.

The alternate mechanism is called SN2 (substitution, nucleophilic, bimolecular) because two molecules are involved in the rate determining (and only) step. Such reactions exhibit overall second-order kinetics. The rate is proportional to both the concentration of the substrate and the concentration of the nucleophile. Reactions like this will have a rate law in the form "rate = k[substrate] [nucleophile]".

Free Radicals Subsitution Reactions

These are the reactions in which an atom or group of atoms in a molecule is replaced by a free radical. The replacement of H- atom by a halogen atom is an example of free radicals substitution. In the free radical substitution reaction, the attacking reagent is free radicals. These reactions are carried either at high temperature or in the presence of ultra-violet light.

In organic chemistry, a radical substitution reaction is a substitution reaction involving free radicals as a reactive intermediate. The reaction always involves at least two steps, and possibly a third.

In the first step called initiation (2,3) a free radical is created by photolysis. Homolysis can be brought about by heat or light but also by radical initiators such as organic peroxides or azo compounds. Light is used to create two free radicals from one diatomic species. The final step is called termination (6,7) in which the radical recombines with another radical species. If the reaction is not terminated, but instead the radical group(s) go on to react further, the steps where new radicals are formed and then react is collectively known as propagation (4,5) because a new radical is created available for secondary reactions.

Mechanism: The mechanism of free radicals substitution involves three steps:-

- Initiation: In this step, halogen molecule breaks homolyticallly to form free radicals. Consider the action of Br_2 on ethane in presence of sunlight.

  Br-Br Br

- Propagation step: The Br formed in the first step reacts with alkane molecule to form new free radicals which in turn reacts with bromine molecule and the chain react ion starts and so on.

  $CH_3CH_3 + Br \rightarrow CH_3\ CH_2 + HBr$

  $CH_3\ CH_2 + Br\text{-}Br \rightarrow CH_3\ CH_2Br + Br$

- Termination: In this step, the free radicals combine and the reaction stops.

  $Br + Br \rightarrow Br_2$

Similarly, consider action of Cl_2 on propene.

- cl – cl→ 2Cl
- $CH_2 = CH\text{-}\ CH_3 + Cl \rightarrow CH_2 = CH - CH_2 + HCl$

  $CH_2{=}CH\text{-}\ CH_2 + Cl - Cl \rightarrow CH_2{=}CH - CH_2\ Cl + Cl$

Substitution Reactions

In an acid-base reaction such as $CH_3CO_2H + NH_3 \rightarrow CH_3CO_2 - + NH_4+$ the N acts as a nucleophile (Greek for "loving the nucleus), the H acts as an electrophile ("loves electrons"), and the O that accepts the pair of electrons acts as a leaving group. The acid-base reaction is the simplest model for a substitution reaction, which is a reaction in which a σ bond between atom 1 and atom 2 is replaced by a σ bond between atom 1 and atom 3. Substitution reactions are incredibly important in organic chemistry, and the most important of these involve substitutions at C. For example:

This substitution reaction, discovered in 1849, involves the nucleophilic O making a new bond to the electrophilic C, and the bond between the electrophilic C and the leaving group I breaking. Any Brønsted base can also act as a nucleophile, and any nucleophile can also act as a Brønsted base, but some compounds are particularly good bases and particularly poor nucleophiles, whereas some are particularly poor bases and particularly good nucleophiles. Any Brønsted or Lewis acid can also act as an electrophile, but there are many electrophiles that are neither Brønsted nor Lewis acids. A haloalkane, e.g. CH_3CH_2Br, can in principle undergo either of two polar reactions when it encounters a lone pair nucleophile, e.g. MeO-. First, MeO- might replace Br- at the electrophilic C atom, forming a new C-O bond and giving ether as the product. This is substitution, because the C-Br σ bond is replaced with a C-O σ bond. Second, MeO- might attack a H atom that is adjacent to the electrophilic C atom, giving MeOH, Br-, and an alkene as products. The electrons in the C-H bond move to form the π bond, and the electrons in the C-X bond leave with X-. This is elimination, because a new π bond is formed, and because the elements of the organic starting material are now divided between more than one product. Elimination requires that the substrate have a C-X bond and adjacent C-H bonds, while substitution requires only that the substrate have a C-X bond.

Nucleophilic Aromatic Substitution Reaction

The name in the title in given to those substitution in which nucleophilic reagent, such as Br^-, combine with aromatic carbon and previously present substituent such as •Cl, •NO_2, becomes expelled along with its bonding electrons with considerable difficultly even •H may be expelled with its bonding electrons i.e at H^-. Biomolecular substitution reaction electron attracting substituent especially one conjugated with aromatic system such as nitro, carbonyl, syano aids the attack of the reagent and a 2-or 4-situated hetero atom, as n pyridine, acts in a similar way. Neuclophilic aromatic substitution can proceed by several, mechanism. The Unimolecular and Biomolecular mechanism can definitely be recognized and other mechanism some of which are understood, can be seen to exist. The unimolecular mechanism is limited to the replacement of those substituents which are sufficiently loosely bound to undergo spontaneous heterolysis in solution. The biomolecular mechanism is much more general, doubtless because it make much less severe demands on the quality of the expelled group, so that a hydrogen shift involved.

Elimination Reaction

Elimination reactions are endothermic reactions which occur at high temperature (>50 °C). Saturated alkyl halides and alcohols give unsaturated alkenes by elimination reactions. In elimination reactions competition occurs between nucleophilic substitution and elimination. Elimination reactions are used in the preparation of alkenes by alkyl halides and alcohols.

- α - Carbon and α - Hydrogen: α – carbon refers to the 1st carbon atom attached to a functional group in an organic molecule. Carbon atom directly attached to a halogen group, aromatic group or double or triple bond is also called α – carbon. All the hydrogen atoms attached to the α – carbon are called α - hydrogens.

- β - Carbon and β - Hydrogen: The carbon atom present next to – carbon is called – carbon and the hydrogen atoms attached to β- carbons are called β - hydrogens.

An elimination reaction is a type of organic reaction in which two substituents are removed from a molecule in either a one or two step mechanism.

Types of Elimination Reactions

Elimination reactions can be of following three types on the basis of mechanisms:

- E1 Reaction,
- E2 Reaction,
- E1 CB (Conjugate Base) reaction.

All three types of reactions differ in their mechanisms. Although all are endothermic reactions and occur at high temperature. In all three reactions the product is always an alkene.

Addition Reaction

Addition reaction occurs when two or more reactants combine to form a product without the loss of any atoms present in the reactants. Addition reaction is common in compounds that have unsaturated C-C bond, like double (alkene) and triple (alkyne) bonds. The weaker π bond is converted into two new stronger σ bonds. Alkenes can give rise to a wide variety of products with various region-selectivity and stereo-selectivity depending on the reagent and reaction conditions. There are typically two types of addition reactions – electrophilic addition and nucleophilic addition. Michael Addition is a type of nucleophilic addition reaction.

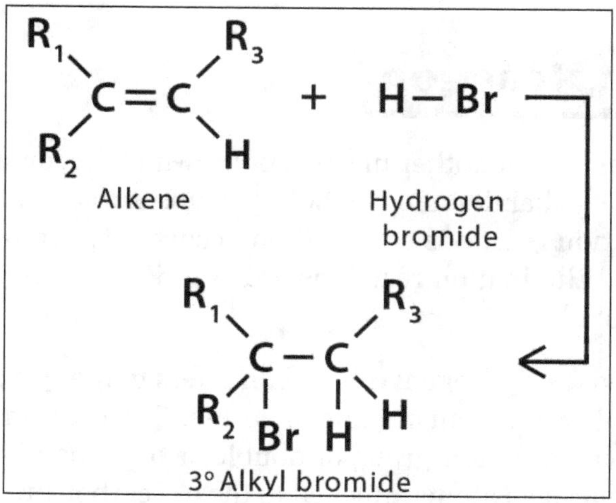

Examples of Addition Reaction

Examples of addition reactions that are common in alkene include halogenation, hydration, hydro-halogenation, hydrogenation, and polymerization.

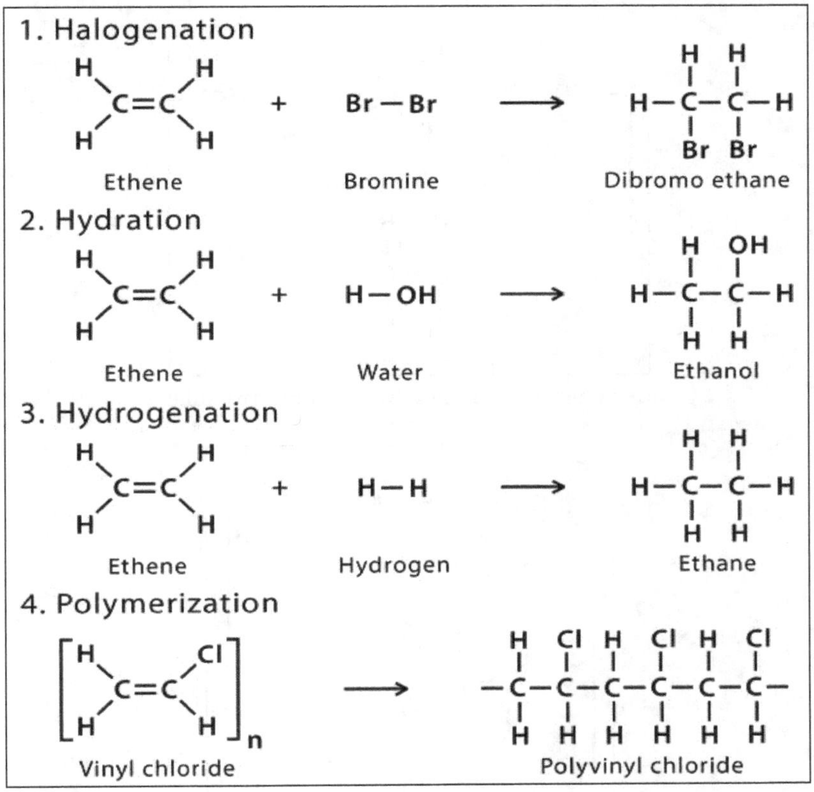

Alkyne can undergo addition reaction to give an alkene or its derivative.

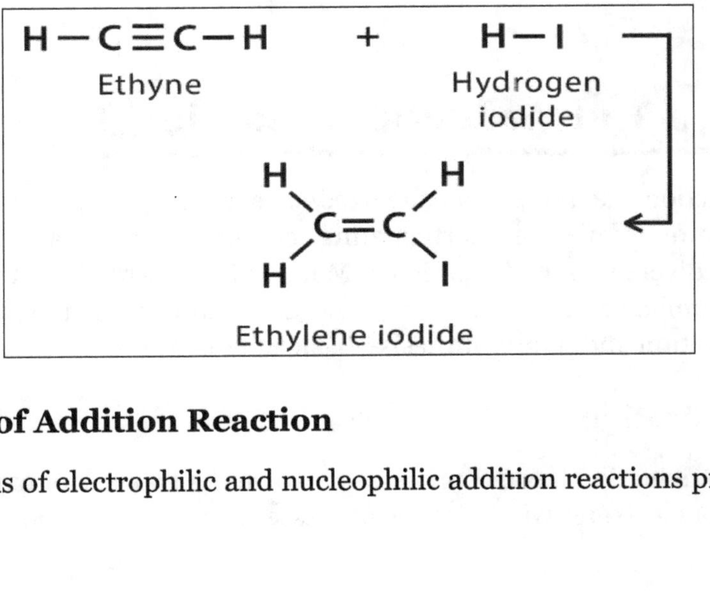

Mechanism of Addition Reaction

The mechanisms of electrophilic and nucleophilic addition reactions proceed through intermediates.

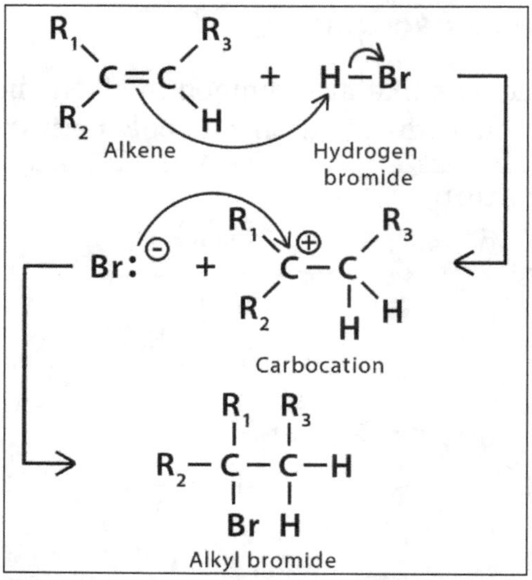

Mechanism of Addition Reaction Electrophilic.

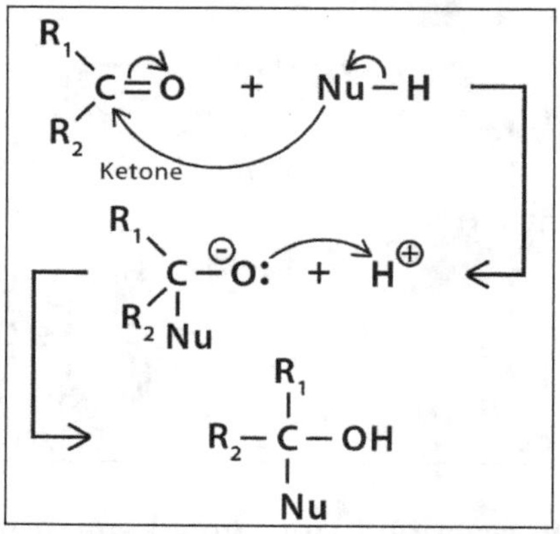

Mechanism of Addition Reaction Nucleophilic.

Oxidation and Reduction Reaction

Oxidation-reduction reaction, also called redox reaction, is any chemical reaction in which the oxidation number of a participating chemical species changes. The term covers a large and diverse body of processes. Many oxidation-reduction reactions are as common and familiar as fire, the rusting and dissolution of metals, the browning of fruit, and respiration and photosynthesis—basic life functions.

Major Classifications

Most oxidation-reduction (redox) processes involve the transfer of oxygen atoms, hydrogen atoms, or electrons, with all three processes sharing two important characteristics: (1) they are coupled—i.e., in any oxidation reaction a reciprocal reduction occurs, and (2) they involve a characteristic net chemical change—i.e., an atom or electron goes from one unit of matter to another. Both reciprocity and net change are illustrated below in examples of the three most common types of oxidation-reduction reactions.

Oxygen-Atom Transfer

Carbon reacts with mercury(II) oxide to produce carbon dioxide and mercury metal. This reaction can be written in equation form:

$$\underset{\text{carbon}}{C} + \underset{\substack{\text{mercury(II)}\\ \text{oxide}}}{2HgO} \longrightarrow \underset{\substack{\text{carbon}\\ \text{dioxide}}}{CO_2} + \underset{\substack{\text{mercury}\\ \text{metal}}}{Hg}$$

Carbon, receiving oxygen, is oxidized; mercury(II) oxide, losing oxygen, undergoes the complementary reduction; and the net change is the transfer of two oxygen atoms from mercury(II) oxide units to a carbon atom.

Hydrogen-Atom Transfer

Hydrogen atoms are transferred from hydrazine, a compound of nitrogen and hydrogen, to oxygen in the following reaction:

$$\underset{\text{hydrazine}}{N_2H_4} + \underset{\text{oxygen}}{O_2} \longrightarrow \underset{\text{nitrogen}}{N_2} + \underset{\text{water}}{2H_2O}$$

Hydrazine, losing hydrogen, is oxidized to molecular nitrogen, while oxygen, gaining hydrogen, is reduced to water.

Electron Transfer

Zinc metal and copper(II) ion react in water solution, producing copper metal and an aqueous (denoted by aq) zinc ion according to the equation:

$$\underset{\substack{\text{zino}\\ \text{metal}}}{Zn} + \underset{\substack{\text{aqueous}\\ \text{copper(II)}\\ \text{ion}}}{Cu^{2+}(aq)} \longrightarrow \underset{\substack{\text{aqueous}\\ \text{zinc(II)}\\ \text{ion}}}{Zn^{2+}(aq)} + \underset{\substack{\text{copper}\\ \text{metal}}}{Cu}$$

With the transfer of two of its electrons, the zinc metal is oxidized, becoming an aqueous zinc ion, while the copper(II) ion, gaining electrons, is reduced to copper metal. Net change is the transfer of two electrons, lost by zinc and acquired by copper.

Because of their complementary nature, the oxidation and reduction processes together are referred to as redox reactions. The reactant that brings about the oxidation is called the oxidizing agent, and that reagent is itself reduced by the reducing agent. In the examples given above, mercury(II) oxide, oxygen, and the copper(II) ion are oxidizing agents, and carbon, hydrazine, and zinc are the reducing agents.

Stoichiometric Basis

Describing the redox processes as above conveys no information about the mechanism by which change takes place. A complete description of the net chemical change for a process is known as the stoichiometry of the reaction, which provides the characteristic combining proportions of elements and compounds. Reactions are classified as redox and non-redox on the basis of stoichiometry; oxygen-atom, hydrogen-atom, and electron transfer are stoichiometric categories.

Oxidation-State Change

Comprehensive definitions of oxidation and reduction have been made possible by modern molecular structure theory. Every atom consists of a positive nucleus, surrounded by negative electrons, which determine the bonding characteristics of each element. In forming chemical bonds, atoms donate, acquire, or share electrons. This makes it possible to assign every atom an oxidation number, which specifies the number of its electrons that can be involved in forming bonds with other atoms. From the particular atoms in a molecule and their known bonding capacities, the bonding pattern within a molecule is determined, and each atom is regarded as being in a specific oxidation state, expressed by an oxidation number.

Redox processes are defined as reactions accompanied by oxidation-state changes: an increase in an atom's oxidation number corresponds to an oxidation; a decrease, to a reduction. In this generalized theory, three examples of ways in which oxidation-state changes can occur are by oxygen-atom (gain, oxidation; loss, reduction), hydrogen-atom (loss, oxidation; gain, reduction), and electron (loss, oxidation; gain, reduction) transfer. The oxidation-state change definition is usually compatible with the above rules for applying the oxygen-atom-transfer and hydrogen-atom-transfer criteria and always compatible with the electron-transfer criterion when it is applicable. The oxidation state of any atom is indicated by a roman numeral following the name or symbol for the element. Thus, iron(III), or Fe(III), means iron in an oxidation state of +3. The uncombined Fe(III) ion is simply Fe^{3+}.

The Redox Concept

Of the chemical processes now regarded as redox reactions, combustion was the earliest focus of philosophical and scientific attention. The Greek scientific philosopher Empedocles listed fire as one of the four elements of matter. In more modern times the phlogiston theory enjoyed scientific popularity. This theory was first articulated in 1697 by German chemist G.E. Stahl. As noted earlier, it asserted that matter releases an elementary constituent, phlogiston, during combustion. Thus, the burning of charcoal was interpreted as the loss of phlogiston from carbon to the air. The theory was also applied to processes other than combustion; in the recovery of a metal from its oxide by heating with charcoal, for example, phlogiston was regarded as being transferred from carbon to the oxide.

Phlogiston saturation was believed to be responsible for the limited ability of air in a closed container to support combustion. A notable consequence of the phlogiston theory was the notion that an oxide of a metal, such as mercury(II) oxide (HgO), was a chemically simpler substance than the metal itself: the metal could be obtained from the oxide only by the addition of phlogiston. The phlogiston theory, however, could provide no acceptable explanation of the gain in weight when an oxide is formed from a metal.

Combustion and Oxide Formation

Late in the 18th century the interrelated work of English chemist Joseph Priestley and French chemist Antoine-Laurent Lavoisier led to the overthrow of the phlogiston theory. Lavoisier saw Priestley's discovery of oxygen in 1774 as the key to the weight gains known to accompany the burning of sulfur and phosphorus and the calcination of metals (oxide formation). In his Traité élémentaire de chimie, he clearly established that combustion consists of a chemical combination between oxygen from the atmosphere and combustible matter. By the end of the century, his ideas were widely accepted and had been successfully applied to the more complex processes of respiration and photosynthesis. Reactions in which oxygen was consumed were classified as oxidations, while those in which oxygen was lost were termed reductions.

Electrochemical Reactions

During the 19th century, the evolving field of electrochemistry led to a broadened view of oxidation. It was possible, for instance, to produce the ferric, or iron(III), ion from the ferrous, or iron(II), ion at the anode (positive electrode, where electrons are absorbed from solution) of an electrochemical cell (a device in which chemical energy is converted to electrical energy), according to the equation:

$$\underset{\text{iron(II) ion}}{Fe^{2+}(aq)} \longrightarrow \underset{\text{iron(II) ion}}{Fe^{3+}(aq)} + \underset{\text{electron}}{e^{-}}$$

Molecular oxygen could effect a similar transformation, according to the equation:

$$\underset{\text{iron(II) ion}}{4Fe^{2+}(aq)} + \underset{\text{oxygen}}{O_2} + \underset{\text{hydrogen ion}}{4H^{+}} \longrightarrow \underset{\text{iron(III) ion}}{4Fe^{3+}(aq)} + \underset{\text{water}}{2H_2O}$$

The similarity of the two processes led to a precursor of the electron-transfer explanation for redox reactions. After the discovery of the electron, the conviction that oxidation and reduction are accomplished through electron loss and gain became firmly entrenched. Thus, early in the 20th century chemists tended to attribute all redox reactions to the transfer of electrons. Later work on chemical bonding, however, demonstrated the incorrectness of that description. An electronegativity scale (listing of elements in descending order of their tendency to attract and hold bonding electrons) provided a firm basis for the oxidation-state assignments on which oxidation-reduction definitions have become based.

Examples of Oxidation-Reduction Reactions

Molecular oxygen is a conspicuously important oxidizing agent. It will directly oxidize all but a few of the metals and most of the nonmetals as well. Often these direct oxidations lead to normal oxides, such as those of lithium (Li), zinc (Zn), phosphorus (P), and sulfur (S).

$$4Li + O_2 \longrightarrow 2LiO_2\,(\text{lithium oxide})$$

$$2Zn + O_2 \longrightarrow 2ZnO\,(\text{zinc oxide})$$

$$4P + 5O_2 \longrightarrow P_4O_{10}\,(\text{phosphorus pentoxide})$$

$$S + O_2 \longrightarrow SO_2\,(\text{sulfur dioxide})$$

Organic foodstuffs are oxidized to carbon dioxide and water in respiration. The reaction stoichiometry can be illustrated for glucose, a simple sugar:

$$\underset{\text{glucose}}{C_6H_{12}O_6} + \underset{\text{oxygen}}{6\,O_2} \longrightarrow \underset{\text{carbon dioxide}}{6CO_2} + \underset{\text{water}}{6H_2O}$$

Although the oxygen-glucose reaction is slow at ambient temperatures outside the living cell, it proceeds quickly under the influence of enzymatic catalysis within the body. Essentially all organic compounds react with oxygen under appropriate conditions, but the reaction rates at ordinary temperatures and pressures vary greatly.

Many other oxidizing agents serve as oxygen-atom sources. Hydrogen peroxide (H_2O_2), acid chromate ion ($HCrO_4^-$), and hypochlorous acid (HClO) are reagents often used in oxygen-atom-transfer reactions—for example, in the following reactions:

$$H_2O_2 + Cl^- \longrightarrow OCl^- + H_2O$$

$$\underset{\text{acid chromate ion}}{2HCrO_4^-} + \underset{\text{acetaldehyde}}{3H_3C\overset{\overset{O}{\|}}{C}H} + \underset{\text{hydrogen ion}}{8H^+} \longrightarrow$$

$$\underset{\text{acetic acid}}{3H_3C\overset{\overset{O}{\|}}{C}OH} + \underset{\text{chromium(III) ion}}{2Cr^{3+}(aq)} + \underset{\text{water}}{5H_2O}$$

$$\underset{\text{hypochlorous acid}}{5HClO} + \underset{\text{iodine}}{I_2} + \underset{\text{water}}{H_2O} \longrightarrow \underset{\text{hydrogen chloride}}{5\,HCl} + \underset{\text{iodic salt}}{2HIO_3}$$

In the simplest hydrogen-atom transfers, molecular hydrogen serves as the hydrogen-atom source. The hydrogenations of ethylene and of molecular nitrogen are illustrative in the following equations:

$$\underset{\text{ethylene}}{H_2C{=}CH_2} + \underset{\text{hydrogen}}{H_2} \longrightarrow \underset{\text{ethane}}{H_3CCH_3}$$

$$\underset{\text{nitrogen}}{N_2} + \underset{\text{hydrogen}}{3H_2} \longrightarrow \underset{\text{ammonia}}{2NH_3}$$

Reactions of molecular hydrogen are characteristically slow at ordinary temperatures. The hydrogenation of molecular nitrogen and of olefins such as ethylene (an olefin is an unsaturated hydrocarbon compound; it has at least two adjacent carbon atoms joined by a double bond to which other atoms or groups of atoms can be joined directly) is a process of extraordinary commercial importance and requires catalysts to occur at useful rates.

Hydrogen-atom transfer from an organic molecule to a suitable acceptor is a common mode of organic oxidation. The oxidation of formic acid by permanganate and that of ethanol by acid chromate share stoichiometry that features hydrogen-atom loss by the organic species, as shown in the following equations:

$$\underset{\text{formic acid}}{3H\overset{\overset{O}{\|}}{C}OH} + \underset{\text{permanganate ion}}{2MnO_4^-} + \underset{\text{hydrogen ion}}{2H^+} \longrightarrow$$

$$\underset{\text{carbon dioxide}}{3CO_2} + \underset{\text{maganese dioxide}}{2MnO_2} + \underset{\text{water}}{4H_2O}$$

$$\underset{\text{ethanol}}{3CH_3CH_2OH} + \underset{\text{acid chromate ion}}{2HCrO_4^-} + \underset{\text{hydrogen ion}}{8H^+} \longrightarrow$$

$$\underset{\text{water}}{8H_2O} + \underset{\text{acetaldehyde}}{3CH_3CHO} + \underset{\text{aqueous chromium(III) ion}}{2Cr^{3+}(aq)}$$

The oxidizing agents permanganate and acid chromate, typical of many hydrogen-atom acceptors, undergo complicated changes rather than simple hydrogen-atom addition. Electron-transfer stoichiometry is usually associated with metal ions in aqueous solution, as shown in the following equations:

$$\underset{\text{iron(II) ion}}{Fe^{2+}(aq)} + \underset{\text{cobalt(III) ion}}{Co^{3+}(aq)} \longrightarrow \underset{\text{iron(III) ion}}{Fe^{3+}(aq)} + \underset{\text{cobalt(II) ion}}{Co^{2+}(aq)}$$

$$\underset{\text{thallium(I) ion}}{Tl^{+}(aq)} + \underset{\text{cerium(IV) ion}}{2Ce^{4+}(aq)} \longrightarrow \underset{\text{thallium(III) ion}}{Tl^{3+}(aq)} + \underset{\text{cerium(III) ion}}{2Ce^{3+}(aq)}$$

Many positively charged metal ions have been shown to be bonded to water molecules, so that their electron-transfer reaction occurs between rather complex molecular groups. The iron ion formulas above, for example, are more properly written as $[Fe(H_2O)_6]^{2+}$ and $[Fe(H_2O)_6]^{3+}$ to reflect the presence of six water molecules bonded to the metal ion. Simple electron transfer between free ions is known only in the gas phase, as in this argon-sodium reaction:

$$\underset{\text{argon ion}}{Ar^+(g)} + \underset{\text{sodium}}{Na(g)} \longrightarrow \underset{\text{sodium ion}}{Na^+(g)} + \underset{\text{argon}}{Ar(g)}$$

Several other types of redox reactions do not fall in the oxygen-atom, hydrogen-atom, or electron-transfer categories. Among these are reactions of fluorine, chlorine, bromine, and iodine. These four elements, known as the halogens, form diatomic molecules, which are versatile oxidizing agents. The following examples are typical:

$$\underset{\text{fluorine}}{3F_2} + \underset{\text{sulfur}}{S} \longrightarrow \underset{\text{sulfur hexafluoride}}{SF_6}$$

$$\underset{\text{chlorine}}{Cl_2} + \underset{\text{iron(II) ion}}{2Fe^{2+}(aq)} \longrightarrow \underset{\text{chloride ion}}{2Cl^-} + \underset{\text{iron(III) ion}}{2Fe^{3+}(aq)}$$

$$\underset{\text{bromine}}{Br_2} + \underset{\text{ethylene}}{H_2C=CH_2} \longrightarrow \underset{\text{ethylene dibromide}}{BrCH_2CH_2Br}$$

$$\underset{\text{iodine}}{I_2} + \underset{\text{thiosulfate ion}}{2S_2O_3^{2-}} \longrightarrow \underset{\text{iodine ion}}{2I^-} + \underset{\text{polythionate ion}}{S_4O_6^{2-}}$$

Such reactions often qualify as redox processes only in the broad sense that oxidation-state changes occur. The oxidation-state characterization extends oxidation-reduction chemistry to include examples from the reactions of all the chemical elements.

Significance of Redox Reactions

Oxidation-reduction reactions have vast importance not only in chemistry but in geology and biology as well. The surface of Earth is a redox boundary between the planet's reduced metallic core and an oxidizing atmosphere. Earth's crust is largely composed of metal oxides, and the oceans are filled with water, an oxide of hydrogen. The tendency of nearly all surface materials to be oxidized by the atmosphere is reversed by the life process of photosynthesis. Because they are constantly renewed by the photosynthetic reduction of carbon dioxide, life's complex compounds can continue to exist on Earth's surface.

For similar reasons, much of chemical technology hinges on the reduction of materials to oxidation states lower than those that occur in nature. Such basic chemical products as ammonia, hydrogen, and nearly all the metals are produced by reductive industrial processes. When not used as structural materials, these products are re-oxidized in their commercial applications. The weathering of materials, including wood, metals, and plastics, is oxidative, since, as the products of technological or photosynthetic reductions, they are in oxidation states lower than those stable in the atmosphere.

Solar radiation is converted to useful energy by a redox cycle that operates continually on a global scale. Photosynthesis converts radiant energy into chemical potential energy by reducing carbon compounds to low oxidation states, and this chemical energy is recovered either through enzymatic oxidations at ambient temperatures or during combustion at elevated temperatures.

Oxidation States

The idea of assigning an oxidation state to each of the atoms in a molecule evolved from the electron-pair concept of the chemical bond. Atoms within a molecule are held together by the force of attraction that the nuclei of two or more of them exert on electrons in the space between them. In many cases this sharing of electrons can be regarded as involving electron-pair bonds between adjacent nuclei. Electron-pair bonding is often diagrammed so as to show all the bonding and nonbonding valence electrons—e.g., the structures of atomic hydrogen, atomic chlorine, and hydrogen chloride shown below (each dot represents one valence electron):

$$\underset{\text{hydrogen atom}}{H\cdot} + \underset{\text{chlorine atom}}{:\underset{\cdot\cdot}{\overset{\cdot\cdot}{Cl}}\cdot} \longrightarrow \underset{\text{hydrogen chloride molecule}}{H:\underset{\cdot\cdot}{\overset{\cdot\cdot}{Cl}}:}$$

The hydrogen chloride diagram reflects the presence, in the inter-nuclear region, of two electrons that are under the mutual attractive influence of both the hydrogen and chlorine nuclei. Oxidation states for the hydrogen and chlorine in HCl are assigned according to the net charges that remain on H and Cl when the shared electrons are assigned to the atom that has the greater attraction for them. Through physical measurements on isolated atoms and simple molecules, these relative attractive powers have been determined.

The table lists the electronegativity values for some important elements.

Pauling electro-negativities of selected elements	
fluorine	3.98
oxygen	3.44
chlorine	3.16
nitrogen	3.04
bromine	2.96

iodine	2.66
sulfur	2.58
carbon	2.55
hydrogen	2.2
phosphorus	2.19
iron	1.83
sodium	0.93

In the hydrogen chloride molecule the chlorine is more electronegative than hydrogen and is, therefore, assigned both shared electrons. Chlorine has seven valence electrons in its neutral state. Having acquired an eighth electron in its reaction with hydrogen, chlorine is considered to have an oxidation state of −1. Hydrogen, on the other hand, is assigned +1, having lost the single valence electron that it has in its neutral state. Charges arrived at in this way are the basis for oxidation-state assignments, conventionally represented by roman numerals, such as in H(I) and Cl(−I) for the constituents of HCl. Because determination of oxidation states is simply a method of conceptually distributing shared electrons to individual atoms, the same number of electrons must be accounted for, before and after such assignment. The table includes examples of molecules that have multiple bonds. The oxidation states of the atoms involved are added up algebraically in the table, and their sum must always equal the net charge on the molecule. There is no physical reality to oxidation states; they simply represent the results of calculations based on a formal rule.

Table: The oxidation states of the atoms in typical small molecules.

Molecular species	Oxidation-state assignments	Algebraic sums
H : Cl :	H(I), Cl(−I)	1 − 1 = 0
H : O : H	H(I), O(−II)	2(1) − 2 = 0
[: O : Cl :]$^-$	Cl(I), O(−II)	1 − 2 = −1
: O : : C : : O :	C(IV), O(−II)	4 − 2(2) = 0
H H : C : O : H H	C(−II), O(−II), H(I)	4(1) − 2 − 2 = 0
[: O : N : O : : O :]$^-$	N(V), O(−II)	5 − 3(2) = −1
[H H : N : H H]$^+$	H(I), N(−III)	4(1) − 3 = +1

Oxidation states can be assigned for most common molecules with the help of a few guidelines. First, electrons shared by two atoms of the same element are divided equally; accordingly, elements are always in oxidation state of 0, regardless of their allotropic form (allotropic refers to the phenomenon of an element's having two or more forms; e.g., carbon can exist as diamond or graphite and in both cases is in the 0 oxidation state). Second, only fluorine is more electronegative than oxygen. Therefore, except in compounds containing oxygen-oxygen or oxygen-fluorine bonds, oxygen can be reliably assigned the oxidation state −2. Similarly, hydrogen is less electronegative than fluorine, oxygen, nitrogen, chlorine, sulfur, and carbon (F, O, N, Cl, S, and C), so it is in the +1 oxidation state in its combinations with those elements. For many common compounds containing only hydrogen, oxygen, and a third element, the third element's oxidation state can be calculated, assuming oxidation numbers of +1 for hydrogen and −2 for oxygen. When bonds are present between two elements that differ little in electronegativity, however, oxidation-state assignments become doubtful, and the distinction between redox and nonredox processes is not evident.

There is a general reluctance, particularly regarding organic systems, to assume oxidation-state changes when the reaction results can be accounted for by the transfer or addition of water (H_2O), ammonia (NH_3), the hydroxide ion (OH^-), or the ions of hydrogen (H^+), chlorine (Cl^-), bromine (Br^-), or iodine (I^-), or combinations of these species; e.g., the ammonium ion (NH_4^+), hydrogen chloride (HCl). The reason is that, in these molecules and ions, the elements are present in their most typical oxidation states: hydrogen(I), chlorine(−I), oxygen(−II), bromine (−I), iodine(−I), and nitrogen(−III).

Oxygen-Atom Transfer Reactions

The oxidation-state concept clarifies the relationship between oxygen-atom, hydrogen-atom, and electron transfer. The oxygen- and hydrogen-transfer criteria apply only when oxygen and hydrogen occur in their typical oxidation states. An example of an appropriate reaction involving oxygen-atom transfer is the reduction of ferrous oxide by carbon monoxide:

$$\underset{\substack{\text{iron(II)oxide} \\ \text{(ferrous oxide)}}}{FeO} + \underset{\substack{\text{carbon} \\ \text{monoxide}}}{CO} \longrightarrow \underset{\substack{\text{carbon} \\ \text{dioxide}}}{CO_2} + \underset{\text{iron}}{Fe}$$

In terms of oxidation-state changes, this oxygen-atom transfer is equivalent to the two-electron reduction of iron and complementary two-electron oxidation of carbon:

$$Fe(II) + C(II) \longrightarrow C(IV) + Fe(O)$$

Oxygen, which occurs in the oxidation state −2 in both reactants and products in the first equation, is not shown in the second. In transferring, the oxygen atom leaves two electrons behind, causing the reduction of iron, and acquires two electrons from the carbon atom, oxidizing the carbon.

In a similar way, the hydrogenation of ethylene corresponds to a two-electron reduction of the two-carbon skeleton:

$$\underset{\text{hydrogen}}{H_2} + \underset{\text{ethylene}}{H_2C=CH_2} \longrightarrow \underset{\text{ethane}}{H_3CCH_3}$$

$$\underset{\substack{\text{hydrogen in}\\ \text{oxidation state zero}}}{2H(O)} + \underset{\substack{\text{carbon in}\\ \text{oxidation state}-2}}{2C(-II)} \longrightarrow$$

$$\underset{\substack{\text{hydrogen in}\\ \text{oxidation state}+1}}{2H(I)} + \underset{\substack{\text{carbon in}\\ \text{oxidation state}-3}}{2C(-III)}$$

In this example also, the second equation includes only the atoms that change oxidation states: the four hydrogen atoms initially present in ethylene are in the +1 oxidation state in both reactants and products and are therefore omitted. Each of the two neutral hydrogen atoms can be regarded as giving up an electron to, and thereby reducing, one of the carbon atoms. This example also demonstrates clearly that the oxidation that complements the reduction of ethylene is that of the two hydrogen atoms in H_2—namely, from the 0 to the +1 oxidation state. General application of the oxidation-state concept leads to a formal viewpoint toward all redox reactions as electron-transfer reactions.

Half Reactions

One of the basic reasons that the concept of oxidation-reduction reactions helps to correlate chemical knowledge is that a particular oxidation or reduction can often be carried out by a wide variety of oxidizing or reducing agents. Reduction of the iron(III) ion to the iron(II) ion by four different reducing agents provides an example:

$$\underset{\substack{\text{iron(III)}\\ \text{ion}}}{Fe^{3+}(aq)} + \underset{\substack{\text{vanadium(II)}\\ \text{ion}}}{V^{2}(aq)} \longrightarrow \underset{\substack{\text{iron(II)}\\ \text{ion}}}{Fe^{2+}(aq)} + \underset{\substack{\text{vanadium(III)}\\ \text{ion}}}{V^{3+}(aq)}$$

$$\underset{\substack{\text{iron(III)}\\ \text{ion}}}{2Fe^{3+}(aq)} + \underset{\substack{\text{zinc}\\ \text{metal}}}{Zn(s)} \longrightarrow \underset{\substack{\text{iron(II)}\\ \text{ion}}}{2Fe^{2+}(aq)} + \underset{\substack{\text{zinc(II)}\\ \text{ion}}}{Zn^{2+}(aq)}$$

$$\underset{\substack{\text{iron(III)}\\ \text{ion}}}{2Fe^{3+}(aq)} + \underset{\substack{\text{iodine}\\ \text{ion}}}{2I^-} \longrightarrow \underset{\substack{\text{iron(II)}\\ \text{ion}}}{2Fe^{2+}(aq)} + \underset{\text{iodine}}{I_2}$$

$$\underset{\substack{\text{iron(II)}\\ \text{ion}}}{2Fe^{3+}(aq)} + \underset{\text{hydroquinone}}{C_6H_4(OH)_2} \longrightarrow \underset{\substack{\text{iron(II)}\\ \text{ion}}}{2Fe^{2+}(aq)} + \underset{\text{quinone}}{C_6H_4O_2} + \underset{\substack{\text{hydrogen}\\ \text{ion}}}{2H^+}$$

Production of the same change in the aqueous iron(III) ion by different reductants emphasizes the fact that the reduction is a characteristic reaction of the iron system itself, and, therefore, the process may be written without specifying the identity of the reducing agent in the following way:

$$\underset{\substack{\text{iron(III)}\\ \text{ion}}}{Fe^{3+}(aq)} + \underset{\text{electron}}{e^-} \longrightarrow \underset{\substack{\text{iron(II)}\\ \text{ion}}}{Fe^{2+}(aq)}$$

Hypothetical equations of this type are known as half reactions. The symbol e^-, which stands for an electron, serves as a reminder that an unspecified reducing agent is required to bring about the change. Half reactions can be written, equally, for the reducing agents in the four reactions with ferric ion:

$$\underset{\text{vanadium(II) ion}}{V^{2+}(aq)} \longrightarrow \underset{\text{vanadium(III) ion}}{V^{3+}(aq)} + \underset{\text{electron}}{e^-}$$

$$\underset{\text{zinc(O) metal}}{Zn(s)} \longrightarrow \underset{\text{zinc(II) ion}}{Zn^{2+}(aq)} + \underset{\text{electron}}{2e^-}$$

$$\underset{\text{iodine ion}}{2I^-} \longrightarrow \underset{\text{iodine}}{I_2} + \underset{\text{electron}}{2e^-}$$

$$\underset{\text{hydroquinone}}{C_6H_4(OH)_2} \longrightarrow \underset{\text{quinone}}{C_6H_4O_2} + \underset{\text{electron}}{2e^-} + \underset{\text{hydrogen ion}}{2H^+}$$

Although hypothetical, half reactions are properly balanced chemical processes. Since $V^{2+}(aq)$ increases its oxidation number by one, from +2 to +3, in the first half reaction, an electron is shown as a product of the change. Similarly, two electrons are produced when the oxidation number of zinc increases from 0 to +2 in the second half reaction. When half reactions for hypothetical isolated oxidations and reductions are combined, the electrons must cancel if the equation for a possible overall chemical reaction is to result.

The use of half reactions is a natural outgrowth of the application of the electron-transfer concept to redox reactions. Since the oxidation-state principle allows any redox reaction to be analyzed in terms of electron transfer, it follows that all redox reactions can be broken down into a complementary pair of hypothetical half reactions. Electrochemical cells (in which chemical energy can be converted to electrical energy, and vice versa) provide some physical reality to the half-reaction idea. Oxidation and reduction half reactions can be carried out in separate compartments of electrochemical cells, with the electrons flowing through a connecting wire and the circuit completed by some arrangement for ion migration between the two compartments (but the migration need not involve any of the materials of the oxidation-reduction reactions themselves).

Redox Potentials For Common Half Reactions

The analysis of the electrical potential, or voltage, developed by pairing various half reactions in electrochemical cells has led to the determination of redox potentials for a substantial number of common half reactions. While a detailed description of redox potentials requires the methods of thermodynamics (the branch of physics concerned with the role played by heat in the transformation of matter or energy), a great deal of useful information can be obtained from redox potentials with minimal recourse to formal theory. Basically, a table of half-cell potentials is a summary of the relative tendencies of different oxidations and reductions to occur.

The table of standard reduction potentials lists selected half reactions and their corresponding reduction potentials (which are symbolized by E°). The physical significance of the values is directly linked to several agreements about their use. First, the greater the value of E° (the reduction potential), the greater the tendency of a half reaction to proceed from left to right (as written). The half reactions in the table are listed from top to bottom in order of decreasing E°: the higher a reaction's position on the list, the greater the tendency of the reactants to accept electrons. In other words, reagents high on the list, such as fluorine gas (F_2) and permanganate ion (MnO_4^-), are strong oxidizing agents. Second, the reduction of hydrogen ions (H^+) to hydrogen gas (H^2) is arbitrarily assigned the value 0 volts. Half cells with positive reduction potentials involve reactants that are more readily reduced than H^+; conversely, those with negative potentials involve reactants that are more difficult to reduce than hydrogen ions.

With the aid of reduction potentials, it is possible to predict whether a particular oxidation-reduction reaction can occur. The predictions require breaking down the overall reaction into two half reactions of known reduction potentials. For example, if a strip of zinc metal is dipped into a solution containing copper(II) ion, the possibility exists for a redox process, which can be regarded as the sum of the half reactions aqueous zinc ion (Zn^{2+}[aq]) to zinc metal (Zn[s]) and aqueous copper ion (Cu^{2+}[aq]) to copper metal (Cu[s]), as follows:

$$\underset{\text{zinc(0) metal}}{Zn(s)} + \underset{\text{copper(II) ion}}{Cu^{2+}(aq)} \longrightarrow \underset{\text{zinc(II) ion}}{Zn^{2+}(aq)} + \underset{\text{copper(0) metal}}{Cu(s)}.$$

$$Cu^{2+}(aq) + 2e^- \longrightarrow Cu(s) \qquad +0.34$$

$$Zn(s) \longrightarrow Zn^{2+}(aq) + 2e^- \qquad +0.76$$

$$\underset{\text{copper(II) ion}}{Cu^{2+}(aq)} + \underset{\text{zinc metal}}{Zn(s)} \xrightarrow{\text{spontaneous}} \underset{\text{copper metal}}{Cu(s)} + \underset{\text{zinc(II) ion}}{Zn^{2+}(aq)} \qquad +1.10 \text{ volt}$$

The resulting E° value for the net reaction, +1.10 volts, measures the tendency of the net reaction to occur. If Eo for a particular net reaction is positive, the process may be expected to occur spontaneously when the reactants are mixed at specified concentrations. Therefore, it is predicted that copper metal should be deposited on a strip of zinc metal when the latter is immersed in a solution of a copper(II) salt. This reaction is, in fact, readily observed in the laboratory. A more specific physical interpretation of the +1.10 volt value is that it represents the voltage that would be produced by an ideal electrochemical cell based on the copper(II) ion to copper metal and zinc(II) ion to zinc metal half reactions with all the reagents at specified concentrations.

When the same two half cells are combined, with both their directions (and therefore the signs of their redox potentials) reversed, it is predicted that the reverse reaction, the depositing of zinc metal from a zinc(II) ion solution onto a copper strip, will not occur spontaneously. As in the case of E° values for half reactions, those for net redox reactions also change sign when the direction of the reaction is reversed.

$$\begin{array}{lr} Cu(s) \longrightarrow Cu^{2+}(aq) + 2e^- & -0.34 \\ Zn^{2+}(aq) + 2e^- \longrightarrow Zn(s) & -0.76 \\ \hline Zn^{2+}(aq) + Cu(s) \xrightarrow{\text{nonspon tan eous}} Cu^{2+}(aq) + Zn(s) & -1.10 \end{array} \text{ volt}$$

The results of the copper-zinc system can be applied more generally to the half reactions in the table of standard reduction potentials. For example, copper(II) ion in water (Cu^{2+}[aq]) is an oxidant strong enough to force a half reaction lower on the table to proceed spontaneously in the opposite direction of that written. Therefore, not only is copper(II) ion expected to oxidize zinc metal (Zn[s]) to zinc(II) ion (Zn^{2+}[aq]); it is also predicted to oxidize hydrogen gas (H_2[g]) to hydrogen ion (H^+) and sodium metal (Na[s]) to sodium ion (Na^+).

Similarly, fluorine gas (F_2[g]), the strongest oxidant listed in the table of standard reduction potentials, is predicted to oxidize spontaneously the products of all the other half reactions in the table. In contrast, the strongest reducing agent is solid sodium metal (Na[s]), and it is expected spontaneously to reduce the reactants of all the other half cells.

Selected values of standard reduction potentials are given in the table.

Half reactions*									E° (volts)
F_2(g) fluorine(o)	+	$2e^-$ electrons	→	$2F^-$ fluoride(−I) ion					2.87
MnO_4^- permanganate ion	+	$8H^+$ hydrogen(I) ions	+	$5e^-$ electrons	→	Mn^{2+}(aq) manganese(II) ion	+	$4H_2O$ water	1.51
Cl_2(g) chlorine(o)	+	$2e^-$ electrons	→	$2Cl^-$ chloride(−I) ions					1.36
O_2(g) oxygen(o)	+	$4H^+$ hydrogen(I) ions	+	$4e^-$ electrons	→	$2H_2O$ water			1.23
Fe^{3+}(aq) iron(III) ion	+	e^- electron	→	Fe^{2+}(aq) iron(II) ion					0.77
Cu^{2+}(aq) copper(II) ion	+	$2e^-$ electrons	→	Cu(s) copper(o)					0.34
$2H^+$ hydrogen(I) ions	+	$2e^-$ electrons	→	H_2(g) hydrogen(o)					0.00
Zn^{2+}(aq) zinc(II) ion	+	$2e^-$ electrons	→	Zn(s) zinc(o)					−0.76
Na^+ sodium(I) ion	+	e^- electron	→	Na(s) sodium(o)					−2.71

*The identifications in parentheses refer to the physical state of the substance: (g), gas; (aq), hydrated positive ion in water; (s), pure solid.

Oxidation-Reduction Equilibria

In practice many chemical reactions can be carried out in either direction, depending on the conditions. The spontaneous direction predicted for a particular redox reaction by half-cell potentials is appropriate to a standard set of reaction conditions. Specifically, the temperature is assumed to be 25 °C with reagents at specified concentrations. Gases are present at one atmosphere pressure and solutes at one mole per litre (one molecular weight in grams dissolved in one litre of solution) concentration (1M). Solids are assumed to be in contact with the reaction solution in their normal stable forms, and water is always taken to be present as the solvent. Many practical problems can be solved directly with standard reduction potentials.

The usefulness of reduction potentials is greatly extended, however, by a thermodynamic relationship known as the Nernst equation, which makes it possible to calculate changes in half-cell potentials that will be produced by deviations from standard concentration conditions. In the reaction between zinc metal and copper(II) ion, standard conditions for zinc and copper metal require simply that both solids be present in contact with the solution; the E° values are not affected by either the total or proportionate amounts of the two metals. The calculation that the overall reaction is spontaneous by +1.10 volts is based on standard one mole per litre (1M) concentrations for aqueous zinc(II) ion (Zn^{2+}[aq]) and aqueous copper(II) ion (Cu^{2+}[aq]). Using the Nernst equation it is found that E° for the overall reaction will be +1.10 volts as long as both ions are present in equal concentrations, regardless of the concentration level.

On the other hand, if the ratio of the zinc(II) to copper(II) ion concentrations is increased, the reduction potential (E°) falls until, at a very high preponderance of zinc ion, E° becomes 0 volt. At this point, there is no net tendency for the reaction to proceed spontaneously in either direction. If the zinc(II) to copper(II) ion ratio is increased further, the direction of spontaneity reverses, and zinc ion spontaneously oxidizes copper metal. In practice, such high zinc(II) to copper(II) ion concentration ratios are unattainable, which means that the reaction can only be carried out spontaneously with copper(II) ion oxidizing zinc metal. Many reactions with E° values smaller than +1.10 volts under standard conditions can be carried out in either direction by adjusting the ratio of product and reactant concentrations. The point at which E° = 0 volt represents a state of chemical equilibrium. When chemical reactions are at equilibrium, the concentrations of the reagents do not change with time, since net reaction is not spontaneous in either direction. Measurements of half-cell potentials combined with Nernst-equation calculations are a powerful technique for determining the concentration conditions that correspond to chemical equilibrium.

Reaction Rates

Predictability

There are practical limitations on predictions of the direction of spontaneity for a chemical reaction, the most important arising from the problem of reaction rates. An analogy can be made with the simple physical system of a block on a sloping plane. Because of the favourable energy change, the block tends spontaneously to slide down, rather than up, the slope, and, at mechanical equilibrium, it will be at the bottom of the slope, since that is the position of lowest gravitational energy. How rapidly the block slides down is a more complex question, since it depends on the amount and kind of friction present. The direction of spontaneity for a chemical reaction is analogous to the downhill direction for a sliding block, and chemical equilibrium is analogous to the position at the bottom of the slope; the rate at which equilibrium is approached depends on the efficiency of the available reaction processes. Between zinc metal and aqueous copper(II) ion, the reaction proceeds without observable delay, but various other spontaneous redox processes proceed at imperceptibly slow rates under ordinary conditions.

Biological Processes

A particularly significant illustration of the role of mechanisms in determining the rates of redox reactions concerns respiration, the central energy-producing process of life. Foodstuffs that are oxidized by molecular oxygen during respiration are quite unreactive with oxygen before ingestion. Such high-energy foods as grains and sugar can resist the atmosphere indefinitely but are rapidly converted to carbon dioxide and water through combination with oxygen during respiratory metabolism. The situation is exemplified by the behaviour of glucose at ambient temperatures.

Living systems are able to use respiratory oxidation as an energy source only because the same reactions are slow outside the cell. In return for providing an efficient mechanism for the oxidation of foods, the cell gains control over the disposition of the liberated chemical energy.

Examples such as the chemistry of respiration make clear the importance of determining the rates and mechanisms of redox reactions. Often questions are difficult to answer even in regard to relatively simple reactions. It has been pointed out that many redox processes can be categorized as oxygen-atom-, hydrogen-atom-, or electron-transfer processes. These categories describe the net changes that are involved but provide no insight into the mechanisms of the reactions.

Mechanisms of Redox Reactions

Some of the problems associated with formulating descriptions of the mechanisms are illustrated by the reaction between two metal ions that undergo complementary, one-unit changes in oxidation state:

$$M(III) + N(II) \rightarrow M(II) + N(III)$$

There are many different metal ions, designated with the letters M and N, which participate in redox reactions with this basic stoichiometry. To define the possible mechanisms with any precision, it is necessary to identify the groups that are directly bonded to the metals, as well as to specify which particular metals take part in the reactions. One relatively simple example is the chromium(II)-iron(III) reaction. Both chromium and iron aquo ions are surrounded by six water molecules in both the +2 and +3 oxidation states.

$$\underset{\text{hexaaquoiron(III) ion}}{Fe(H_2O)_6^{3+}} + \underset{\text{hexaquochromium(II) ion}}{Cr(H_2O)_6^{2+}} \longrightarrow \underset{\text{hexaaquoiron(II) ion}}{Fe(H_2O)_6^{2+}} + \underset{\text{hexaquochromium(III) ion}}{Cr(H_2O)_6^{3+}}$$

The reaction appears from its stoichiometry to entail simple electron transfer from Cr(II) to Fe(III). There are, however, other possibilities. Their variety is most evident if it is assumed that the oxidizing agent, hexaaquoiron(III) ($Fe\left[H_2O_6^{3+}\right]$), dissociates a hydrogen ion before reacting with hexaaquochromium(II) ($Cr\left[H_2O_6^{2+}\right]$). The hexaaquoiron(III) ion is then converted to a hydroxopentaaquoiron(III) ion, known to be an important reactant in the Fe(III)-Cr(II) reaction. Its formation requires no change in the oxidation state of iron and should be regarded as a simple acid-base reaction. With hydroxopentaaquoiron(III) ion, $(H_2O)_5FeOH^{2+}$, as the oxidizing agent, simple electron transfer does provide one plausible means for the reaction. All the steps in the mechanism must add up to give the overall reaction. Intermediate species such as hydroxopentaaquoiron(III) must cancel out in the addition process by occurring as products in one step and reactants in another:

$$\underset{\text{hexaaquoiron(III) ion}}{Fe(H_2O)_6^{3+}} \longrightarrow \underset{\text{hydroxopenta-aquoiron(III) ion}}{(H_2O)_5FeOH^{2+}} + \underset{\text{hydrogen ion}}{H^+}$$

$$\underset{\text{hydroxopenta-aquoiron(III) ion}}{(H_2O)_5FeOH^{2+}} + \underset{\text{hexaaquo-chromium(III) ion}}{Cr(H_2O)_6^{2+}}$$

$$\xrightarrow[\text{transfer}]{\text{electron}} \underset{\text{hydroxopenta-aquoiron(II) ion}}{(H_2O)_5FeOH^{+}} + \underset{\text{hexaaquo-chromium(III) ion}}{Cr(H_2O)_6^{3+}}$$

$$\underset{\text{hydroxopenta-aquoiron(II) ion}}{(H_2O)_5FeOH^{+}} + \underset{\text{hydrogen ion}}{H^+} \longrightarrow \underset{\text{hexaaquoiron(II) ion}}{Fe(H_2O)_6^{2+}}$$

$$Fe(H_2O)_6^{3+} + Cr(H_2O)_6^{2+} \longrightarrow Fe(H_2O)_6^{2+} + Cr(H_2O)_6^{3+}$$

An alternative to simple electron transfer is the transfer of a hydrogen atom from a water molecule bound to chromium(II) to the hydroxide ion (OH^-) bound to iron(III). A third possibility is the transfer of a neutral OH group (called a hydroxyl radical) from iron(III) to chromium(II). Oxygen and hydrogen occur as oxygen(−II) and hydrogen(I) in both reactants and products. These oxidation states correspond to those present in the hydroxide ion OH^-. Therefore, when an OH group transfers, it must leave an electron behind, reducing the iron, and then reaccept an electron from chromium, oxidizing it. Since both the aqueous iron and chromium ions have a strong tendency to remain surrounded by six oxygen molecules, the most likely method of OH transfer would be the sharing of the transferred oxygen between the two metals at an intermediate stage of the reaction. But iron(III) is not a strong enough oxidizing agent to produce a significant concentration of neutral OH groups from OH^- ions. A more acceptable description of the OH transfer mechanism, therefore, is to regard OH^- as a bridging group through which electron transfer is accomplished as the two metal ion groups share a bonded oxygen atom.

$$\left[(H_2O)_5\overset{3+}{Fe}-\underset{\underset{H}{|}}{O}-\overset{2+}{Cr}(H_2O)_5\right]^{4+} \longrightarrow \left[(H_2O)_5\overset{2+}{Fe}-\underset{\underset{H}{|}}{O}-\overset{3+}{Cr}(H_2O)_5\right]^{4+}$$

Experimentally, it is difficult to distinguish between the alternative mechanisms posed above. Classification schemes for reaction mechanisms are of little value unless they offer alternatives that can be experimentally verified. For that reason the most successful system for classifying the mechanisms of redox reactions between metal ions has been the inner sphere–outer sphere dichotomy. Outer-sphere reactions are those that take place without breaking any bonds between a metal and a group such as water or hydroxide ion bound to it. Both the simple electron- and hydrogen-atom-transfer mechanisms are outer-sphere processes. If, on the other hand, the oxidation-state changes take place within an intermediate product in which the two metals are directly bonded to a common bridging group (such as shown in the equation above), the mechanism is inner-sphere. Formation of the intermediate requires loss of a bound water molecule by one of the metals. The inner sphere–outer sphere distinction can be tested experimentally in some redox systems.

References

- Organic-reaction, encyclopedia: chemeurope.com, Retrieved 02, July 2020
- Substitution-reaction, chemistry: ukessays.com, Retrieved 10, March 2020
- Elimination-reaction, chemistry: vedantu.com, Retrieved 05, August 2020
- Addition-reaction, addition-reaction-occurs-when-two-new-stronger-CF-bonds: chemistrylearner.com, Retrieved 22, March 2020
- Oxidation-reduction-reaction, science: britannica.com, Retrieved 07, April 2020

4

Types of Organic Synthesis

Organic synthesis is concerned with the strategies involved in the intentional construction of organic compounds from easily available starting materials. Main areas of organic synthesis include total synthesis, semi-synthesis and methodology. Science and technology have undergone rapid developments in the past decade which has resulted in the discovery of significant tools and techniques in the field of organic synthesis, which have been extensively detailed in this chapter.

Organic Synthesis

Organic synthesis is a special branch of chemical synthesis and is concerned with the construction of organic compounds via organic reactions. Organic molecules often contain a higher level of complexity than purely inorganic compounds, so that the synthesis of organic compounds has developed into one of the most important branches of organic chemistry. There are several main areas of research within the general area of organic synthesis: *total synthesis*, *semisynthesis*, and *methodology*.

Total Synthesis

A total synthesis is the complete chemical synthesis of complex organic molecules from simple, commercially available (petrochemical) or natural precursors. Total synthesis may be accomplished either via a linear or convergent approach. In a *linear* synthesis—often adequate for simple structures—several steps are performed one after another until the molecule is complete; the chemical compounds made in each step are called *synthetic intermediates*. For more complex molecules, a convergent synthetic approach may be preferable, one that involves individual preparation of several "pieces" (key intermediates), which are then combined to form the desired product.

Robert Burns Woodward, who received the 1965 Nobel Prize for Chemistry for several total syntheses (e.g., his 1954 synthesis of strychnine), is regarded as the father of modern organic synthesis. Some latter-day examples include Wender's, Holton's, Nicolaou's, and Danishefsky's total syntheses of the anti-cancer therapeutic, paclitaxel (trade name, Taxol).

Methodology and Applications

Each step of a synthesis involves a chemical reaction, and reagents and conditions for each of these reactions must be designed to give an adequate yield of pure product, with as little work as possible. A method may already exist in the literature for making one of the early synthetic intermediates, and this method will usually be used rather than an effort to "reinvent the wheel". However, most intermediates are compounds that have never been made before, and these will normally be made using general methods developed by methodology researchers.

To be useful, these methods need to give high yields, and to be reliable for a broad range of substrates. For practical applications, additional hurdles include industrial standards of safety and purity.

Methodology research usually involves three main stages: *discovery*, *optimisation*, and studies of *scope and limitations*. The *discovery* requires extensive knowledge of and experience with chemical reactivities of appropriate reagents. *Optimisation* is a process in which one or two starting compounds are tested in the reaction under a wide variety of conditions of temperature, solvent, reaction time, etc., until the optimum conditions for product yield and purity are found. Finally, the researcher tries to extend the method to a broad range of different starting materials, to find the scope and limitations. Total syntheses are sometimes used to showcase the new methodology and demonstrate its value in a real-world application. Such applications involve major industries focused especially on polymers (and plastics) and pharmaceuticals.

Stereoselective Synthesis

Most complex natural products are chiral, and the bioactivity of chiral molecules varies with the enantiomer. Historically, total syntheses targeted racemic mixtures, mixtures of both possible enantiomers, after which the racemic mixture might then be separated via chiral resolution.

In the later half of the twentieth century, chemists began to develop methods of stereoselective catalysis and kinetic resolution whereby reactions could be directed to produce only one enantiomer rather than a racemic mixture. Early examples include stereoselective hydrogenations (e.g., as reported by William Knowles and Ryōji Noyori), and functional group modifications such as the asymmetric epoxidation of Barry Sharpless; for these specific achievements, these workers were awarded the Nobel Prize in Chemistry in 2001. Such reactions gave chemists a much wider choice of enantiomerically pure molecules to start from, where previously only natural starting materials could be used. Using techniques pioneered by Robert B. Woodward and new developments in synthetic methodology, chemists became more able to take simple molecules through to more complex molecules without unwanted racemisation, by understanding stereocontrol, allowing final target molecules to be synthesised pure enantiomers (i.e., without need for resolution). Such techniques are referred to as *stereoselective synthesis*.

Synthesis Design

Elias James Corey brought a more formal approach to synthesis design, based on retrosynthetic analysis, for which he won the Nobel Prize for Chemistry in 1990. In this approach, the synthesis is planned backwards from the product, using standard rules. The steps "breaking down" the parent structure into achievable component parts are shown in a graphical scheme that uses *retrosynthetic arrows*.

More recently, and less widely accepted, computer programs have been written for designing a synthesis based on sequences of generic "half-reactions".

Retrosynthetic Analysis

Retrosynthetic analysis is a technique for solving problems in the planning of organic syntheses. This is achieved by transforming a target molecule into simpler precursor structures without assumptions regarding starting materials. Each precursor material is examined using the same method. This procedure is repeated until simple or commercially available structures are reached. E.J. Corey formalized this concept in his book *The Logic of Chemical Synthesis*.

The power of retrosynthetic analysis becomes evident in the design of a synthesis. The goal of retrosynthetic analysis is structural simplification. Often, a synthesis will have more than one possible synthetic route. Retrosynthesis is well suited for discovering different synthetic routes and comparing them in a logical and straightfoward fashion. A database may be consulted at each stage of the analysis, to determine whether a component already exists in the literature. In that case, no further exploration of that compound would be required.

Definitions

Disconnection:

A retrosynthetic step involving the breaking of a bond to form two (or more) synthons.

Retron:

A minimal molecular substructure that enables certain transformations.

Retrosynthetic tree:

A directed acyclic graph of several (or all) possible retrosyntheses of a single target.

Synthon:

An idealized molecular fragment. A synthon and the corresponding commercially available synthetic equivalent are shown below:

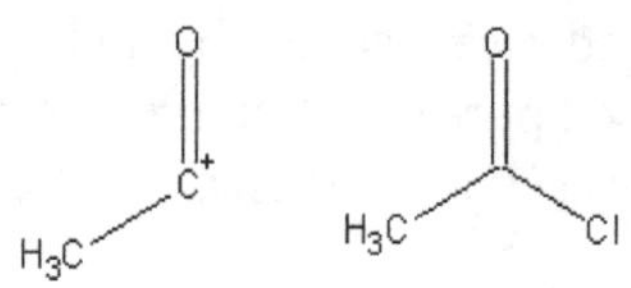

Target:

The desired final compound.

Transform:

The reverse of a synthetic reaction; the formation of starting materials from a single product.

Example

An example will allow the concept of retrosynthetic analysis to be easily understood.

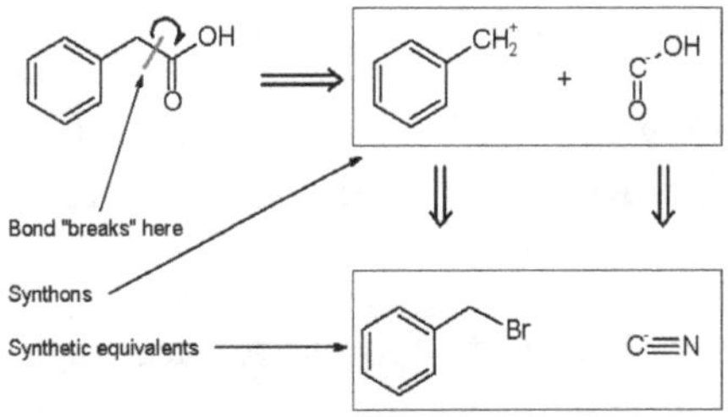

In planning the synthesis of phenylacetic acid, two synthons are identified. A nucleophilic "-COOH" group, and an electrophilic "$PhCH_2^+$" group. Of course, both synthons do not exist per se; synthetic equivalents corresponding to the synthons are reacted to produce the desired product. In this case, the cyanide anion is the synthetic equivalent for the ^-COOH synthon, while benzyl bromide is the synthetic equivalent for the benzyl synthon.

The synthesis of phenylacetic acid determined by retrosynthetic analysis is thus:

$$PhCH_2Br + NaCN \rightarrow PhCH_2CN + NaBr$$

$$PhCH_2CN + 2\ H_2O \rightarrow PhCH_2COOH + NH_3$$

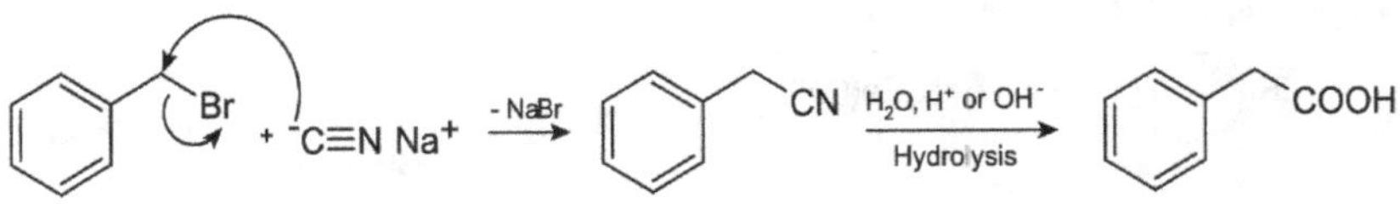

In fact, phenylacetic acid has been synthesized from benzyl cyanide, itself prepared by the analogous reaction of benzyl chloride with sodium cyanide.

Strategies

Functional Group Strategies

Manipulation of functional groups can lead to significant reductions in molecular complexity.

Stereochemical Strategies

Numerous chemical targets have distinct stereochemical demands. Stereochemical transformations (such as the Claisen rearrangement and Mitsunobu reaction) can remove or transfer the desired chirality thus simplifying the target.

Structure-goal Strategies

Directing a synthesis toward a desirable intermediate can greatly narrow the focus of an analysis. This allows bidirectional search techniques.

Transform-based Strategies

The application of transformations to retrosynthetic analysis can lead to powerful reductions in molecular complexity. Unfortunately, powerful transform-based retrons are rarely present in complex molecules, and additional synthetic steps are often needed to establish their presence.

Topological Strategies

The identification one or more key bond disconnections may lead to the identification of key substructures or difficult to identify rearrangement transformations in order to identify the key structures.

- Disconnections that preserve ring structures are encouraged.
- Disconnections that create rings larger than 7 members are discouraged.

Enantioselective Synthesis

In the Sharpless dihydroxylation reaction the chirality of the product can be controlled by the "AD-mix" used. This is an example of enantioselective synthesis using asymmetric induction Key: R_L = Largest substituent; R_M = Medium-sized substituent; R_S = Smallest substituent,

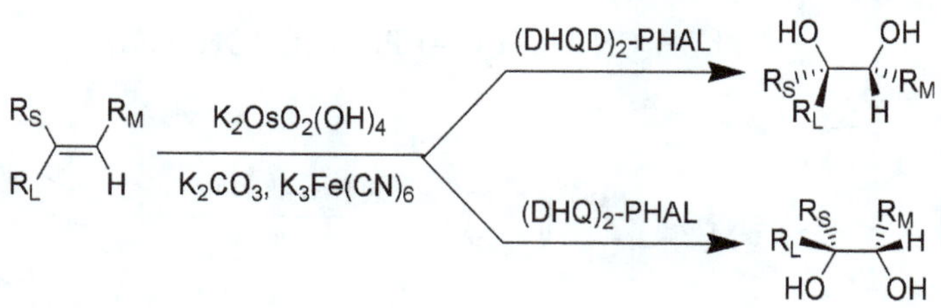

Enantioselective synthesis, also called chiral synthesis or asymmetric synthesis, is a form of chemical synthesis. It is defined by IUPAC as: a chemical reaction (or reaction sequence) in which one or more new elements of chirality are formed in a substrate molecule and which produces the stereoisomeric (enantiomeric or diastereoisomeric) products in unequal amounts.

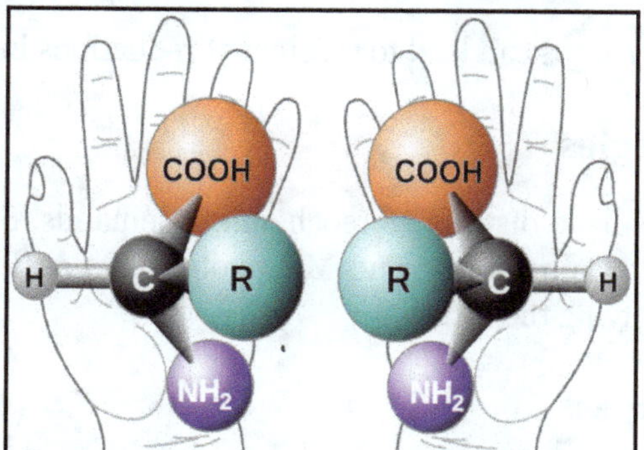

Two enantiomers of a generic alpha amino acid.

Put more simply: it is the synthesis of a compound by a method that favors the formation of a specific enantiomer or diastereomer.

Enantioselective synthesis is a key process in modern chemistry and is particularly important in the field of pharmaceuticals, as the different enantiomers or diastereomers of a molecule often have different biological activity.

Overview

Many of the building blocks of biological systems, such as sugars and amino acids, are produced exclusively as one enantiomer. As a result of this living systems possess a high degree of chemical chirality and will often react differently with the various enantiomers of a given compound. Examples of this selectivity include:

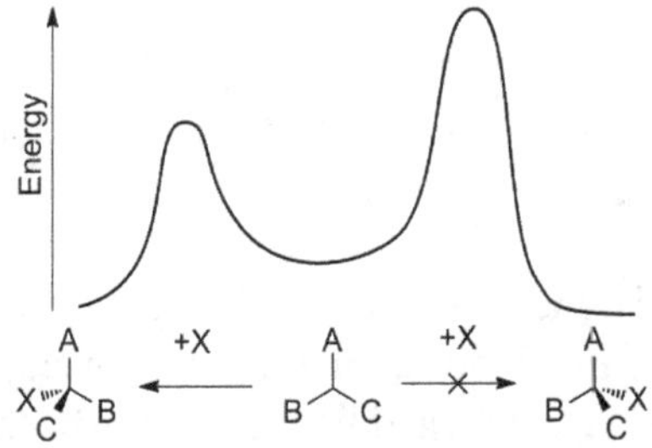

A Gibbs free energy plot of an enantioselective addition reaction.

- Flavour: the artificial sweetener aspartame has two enantiomers. L-aspartame tastes sweet, yet D-aspartame is tasteless.
- Odor: *R*-(−)-carvone smells like spearmint yet *S*-(+)-carvone, smells like caraway.
- Drug effectiveness: the antidepressant drug Citalopram is sold as a racemic mixture. However, studies have shown that only the (*S*)-(+) enantiomer is responsible for the drug's beneficial effects.
- Drug safety: *D*penicillamine is used in chelation therapy and for the treatment of rheumatoid arthritis. However *L*penicillamine is toxic as it inhibits the action of pyridoxine, an essential B vitamin.

As such enantioselective synthesis is of great importance; but it can also be difficult to achieve. Enantiomers possess identical enthalpies and entropies, and hence should be produced in equal amounts by an undirected process – leading to a racemic mixture. The solution is to introduce a chiral feature which will promote the formation of one enantiomer over another via interactions at the transition state. This is known as asymmetric induction and can involve chiral features in the substrate, reagent, catalyst or environment and works by making the activation energy required to form one enantiomer lower than that of the opposing enantiomer.

Asymmetric induction can occur intramolecularly when given a chiral starting material. This behaviour can be exploited, especially when the goal is to make several consecutive chiral centres to give a specific enantiomer of a specific diastereomer. An aldol reaction, for example, is inherently diastereoselective; if the aldehyde is enantiopure, the resulting aldol adduct is diastereomerically and enantiomerically pure.

Approaches

Enantioselective Catalysis

In general, enantioselective catalysis (known traditionally as asymmetric catalysis) refers to the use of chiral coordination complexes as catalysts. This approach is very commonly encountered, as it is effective for a broader range of transformations than any other method of enantioselective synthesis. The catalysts are typically rendered chiral by using chiral ligands, however it is also possible to generate chiral-at-metal complexes using simpler achiral ligands. Most enantioselective catalysts are effective at low concentrations making them well suited to industrial scale synthesis; as even exotic and expensive catalysts can be used affordably. Perhaps the most versatile example of enantioselective synthesis is asymmetric hydrogenation, which is able to reduce a wide variety of functional groups.

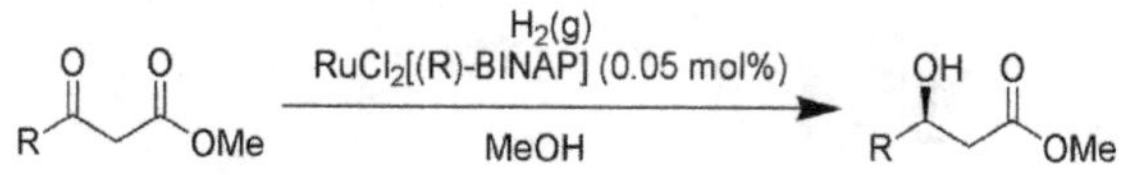

With only 75 natural metals in existence (and not all of these showing extensive catalytic activities) the design of new catalysts is very much dominated by the development of new classes of ligands. Certain ligands, often referred to as 'privileged ligands', have been found to be effective in a wide range of reactions; examples include BINOL, Salen and BOX. However, most catalysts are rarely general, requiring certain functional groups in the substrate to form the transition state complex correctly: arbitrary structures cannot be used. For example, Noyori asymmetric hydrogenation with BINAP/Ru requires a β-ketone, although another catalyst, BINAP/diamine-Ru, widens the scope to α,β-olefins and aromatics.

Chiral Auxiliaries

A chiral auxiliary is an organic compound which couples to the starting material to form new compound which can then undergo enantioselective reactions via intramolecular asymmetric induction. At the end of the reaction the auxiliary is removed, under conditions that will not cause racemization of the product. It is typically then recovered for future use.

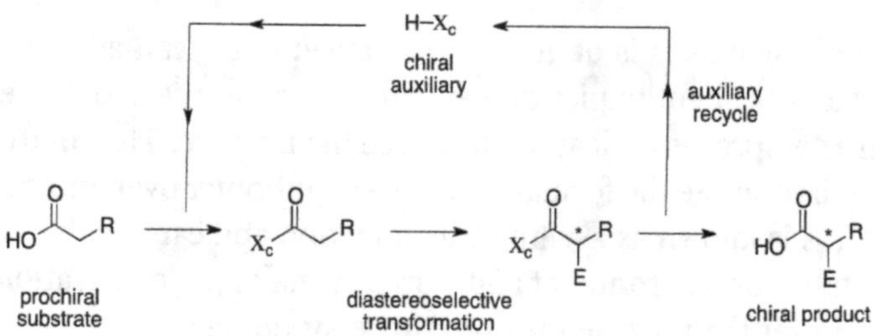

Chiral auxiliaries must be used in stoichiometric amounts to be effective and require additional synthetic steps to append and remove the auxiliary. However, in some cases the only available stereoselective methodology relies on chiral auxiliaries and these reactions tend to be versatile and very well-studied, allowing the most time-efficient access to enantiomerically pure products. Additionally, the products of auxiliary-directed reactions are diastereomers, which enables their facile separation by methods such as column chromatography or crystallization.

Biocatalysis

Biocatalysis makes use of biological compounds, ranging from isolated enzymes to living cells, to perform chemical transformations. The advantages of these reagents include very high ee's and reagent specificity, as well as mild operating conditions and low environmental impact. Biocatalysts are more commonly used in industry than in academic research; for example in the production of statins. The high reagent specificity can be a problem however; as it often requires that a wide range of biocatalysts be screened before an effective reagent is found.

Enantioselective Organocatalysis

Organocatalysis refers to a form of catalysis, where the rate of a chemical reaction is increased by an organic compound consisting of carbon, hydrogen, sulfur and other non-metal elements. When the organocatalyst is chiral enantioselective synthesis can be achieved; for example a number of carbon–carbon bond forming reactions become enantioselective in the presence of proline with the aldol reaction being a prime example. Organocatalysis often employs natural compounds and secondary amines as chiral catalysts; these are inexpensive and environmentally friendly, as no metals are involved.

Chiral Pool Synthesis

Chiral pool synthesis is one of the simplest and oldest approaches for enantioselective synthesis. A readily available chiral starting material is manipulated through successive reactions, often using achiral reagents, to obtain the desired target molecule. This can meet the criteria for enantioselective synthesis when a new chiral species is created, such as in an S_N2 reaction.

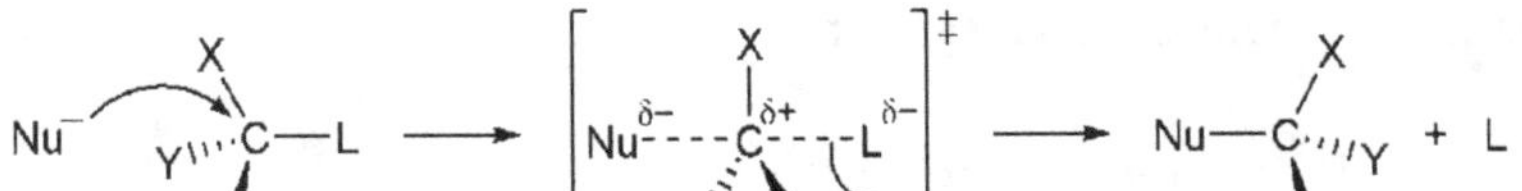

Chiral pool synthesis is especially attractive for target molecules having similar chirality to a relatively inexpensive naturally occurring building-block such as a sugar or amino acid. However, the number of possible reactions the molecule can undergo is restricted and tortuous synthetic routes may be required (e.g. Oseltamivir total synthesis). This approach also requires a stoichiometric amount of the enantiopure starting material, which can be expensive if it is not naturally occurring.

Alternative Approaches

Alternatives to enantioselective synthesis usually involve the isolation of one enantiomer from a racemic mixture by any of a number of methods. If the cost in time and money of making such racemic mixtures is low (or if both enantiomers may find use) then this approach may remain cost-effective. Common methods of separation are based around chiral resolution or kinetic resolution.

Separation and Analysis of Enantiomers

The two enantiomers of a molecule possess the same physical properties (e.g. melting point, boiling point, polarity etc.) and so behave identically to each other. As a result, they will migrate with an identical R_f in thin layer chromatography and have identical retention times in HPLC and GC. Their NMR and IR spectra are identical.

This can make it very difficult to determine whether a process has produced a single enantiomer (and crucially which enantiomer it is) as well as making it hard to separate enantiomers from a reaction which has not been 100% enantioselective. Fortunately, enantiomers behave differently in the presence of other chiral materials and this can be exploited to allow their separation and analysis.

Enantiomers do not migrate identically on chiral chromatographic media, such as quartz or standard media that has been chirally modified. This forms the basis of chiral column chromatography, which can be used on a small scale to allow analysis via GC and HPLC, or on a large scale to separate chirally impure materials. However this process can require large amount of chiral packing material which can be expensive. A common alternative is to use a chiral derivatizing agent to convert the enantiomers into a diastereomers, in much the same way as chiral auxiliaries. These have different physical properties and hence can be separated and analysed using conventional methods. Special chiral derivitizing agents known as 'chiral resolution agents' are used in the NMR spectroscopy of stereoisomers, these typically involve coordination to chiral europium complexes such as $Eu(fod)_3$ and $Eu(hfc)_3$.

The enantiomeric excess of a substance can also be determined using certain optical methods. The oldest method for doing this is to use a polarimeter to compare the level of optical rotation in the product against a 'standard' of known composition. It is also possible to perform ultraviolet-visible spectroscopy of stereoisomers by exploiting the Cotton effect.

One of the most accurate ways of determining the chirality of compound is to determine its absolute configuration by Xray Crystallography. However this is a labour-intensive process which requires that a suitable single crystal be grown.

History

Inception (1815–1905)

In 1815 the French physicist Jean-Baptiste Biot showed that certain chemicals could rotate the plane of a beam of polarised light, a property called optical activity. The nature of this property remained a mystery until 1848, when Louis Pasteur proposed that it had a molecular basis originating from some form of *dissymmetry*, with the term *chirality* being coined by Lord Kelvin a year later. The origin of chirality itself was finally described in 1874, when Jacobus Henricus van 't Hoff and Joseph Le Bel independently proposed the tetrahedral geometry of carbon; structural models prior to this work had been two-dimensional, and van 't Hoff Le Bel theorized that the arrangement of groups around this tetrahedron could dictate the optical activity of the resulting compound.

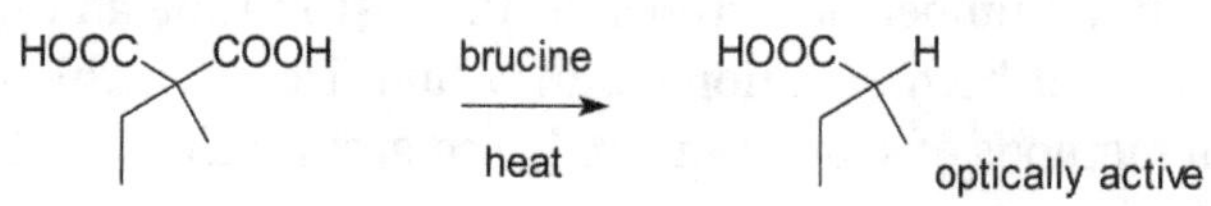

Marckwald's brucine-catalyzed enantioselective decarboxylation of 2-ethyl-2-methylmalonic acid, resulting in a slight excess of the levorotary form of the 2-methylbutyric acid product.

In 1894 Hermann Emil Fischer outlined the concept of asymmetric induction; in which he correctly ascribed selective the formation of D-glucose by plants to be due to the influence of optically active substances within chlorophyll. Fischer also successfully performed what would now be regarded as the first example of enantioselective synthesis, by enantioselectively elongating sugars via a process which would eventually become the Kiliani–Fischer synthesis.

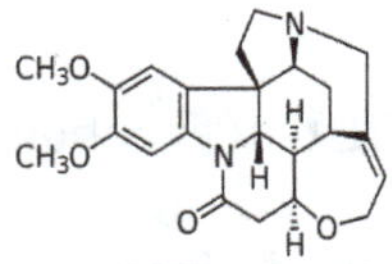

Brucine, an alkaloid natural product related to strychnine, used successfully as an organocatalyst by Marckwald in 1904.

The first enantioselective chemical synthesis is most often attributed to Willy Marckwald, Universität zu Berlin, for a brucine-catalyzed enantioselective decarboxylation of *2-ethyl-2-methylmalonic acid* reported in 1904. A slight excess of the levorotary form of the product of the reaction, 2-methylbutyric acid, was produced; as this product is also a natural product—e.g., as a side chain of lovastatin formed by its diketide synthase (LovF) during its biosynthesis—this result constitutes the first recorded total synthesis with enantioselectivity, as well other firsts (as Koskinen notes, first "example of asymmetric catalysis, enantiotopic selection, and organocatalysis"). This observation is also of historical significance, as at the time enantioselective synthesis could only be understood in terms of vitalism. Natural and artificial compounds were fundamentally different, it was argued, and chirality could only exist in natural compounds. Unlike Fischer, Marckwald had performed an enantioselective reaction upon an achiral, *un-natural* starting material, albeit with a chiral organocatalyst (as we now understand this chemistry).

Early Work (1905–1965)

The development of enantioselective synthesis was initially slow, largely due to the limited range of techniques available for their separation and analysis. Diastereomers possess different physical properties, allowing separation by conventional means, however at the time enantiomers could only be separated by spontaneous resolution (where enantiomers separate upon crystallisation) or kinetic resolution (where one enantiomer is selectively destroyed). The only tool for analysing enantiomers was optical activity using a polarimeter, a method which provides no structural data.

It was not until the 1950s that major progress really began. Driven in part by chemists such as R. B. Woodward and Vladimir Prelog but also by the development of new techniques. The first of these was Xray Crystallography, which was used to determine the absolute configuration of an organic compound by Johannes Bijvoet in 1951. Chiral chromatography was introduced a year later by Dalgliesh, who used paper chromatography to separate chiral amino acids. Although Dalgliesh was not the first to observe such separations, he correctly attributed the separation of enantiomers to differential retention by the chiral cellulose. This was expanded upon in 1960, when Klem and Reed first reported the use of chirally-modified silica gel for chiral HPLC chromatographic separation.

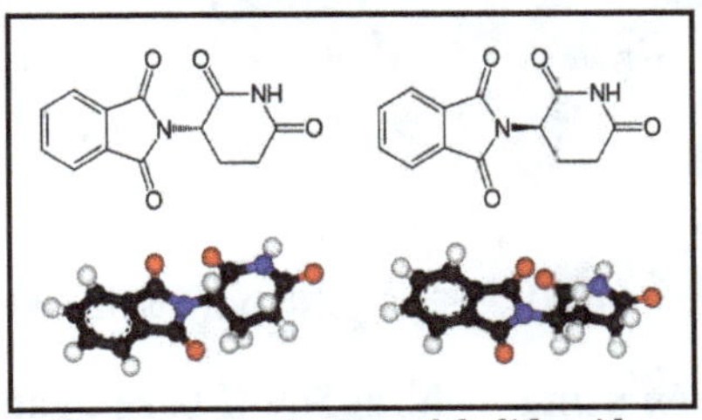

The two enantiomers of thalidomide:
Left: (*S*)-thalidomide.
Right: (*R*)-thalidomide.

Thalidomide

While it had long been known that the different enantiomers of a drug could have different activities, this was not accounted for in early drug design and testing. However following the thalidomide disaster the development and licensing of drugs changed dramatically.

First synthesized in 1953, thalidomide was widely prescribed for morning sickness from 1957 to 1962, but was soon found to be seriously teratogenic, eventually causing birth defects in more than 10,000 babies. The disaster prompted many counties to introduce tougher rules for the testing and licensing of drugs, such as the Kefauver-Harris Amendment (U.S.) and Directive 65/65/EEC1 (E.U.).

Early research into the teratogenic mechanism, using mice, suggested that one enantiomer of thalidomide was teratogenic while the other possessed all the therapeutic activity. This theory was later shown to be incorrect and has now been superseded by a body of research. However it raised the importance of chirality in drug design, leading to increased research into enantioselective synthesis.

Modern Age (Since 1965)

The Cahn–Ingold–Prelog priority rules (often abbreviated as the CIP system) were first published in 1966; allowing enantiomers to be more easily and accurately described. The same year saw first successful enantiomeric separation by gas chromatography an important development as the technology was in common use at the time.

Metal catalysed enantioselective synthesis was pioneered by William S. Knowles, Ryōji Noyori and K. Barry Sharpless; for which they would receive the 2001 Nobel Prize in Chemistry. Knowles and Noyori began with the development of asymmetric hydrogenation, which they developed independently in 1968. Knowles replaced the achiral triphenylphosphine ligands in Wilkinson's catalyst with chiral phosphine ligands. This experimental catalyst was employed in an asymmetric hydrogenation with a modest 15% enantiomeric excess. Knowles was also the first to apply enantioselective metal catalysis to industrial-scale synthesis; while working for the Monsanto Company he developed an enantioselective hydrogenation step for the production of L-DOPA, utilising the DIPAMP ligand.

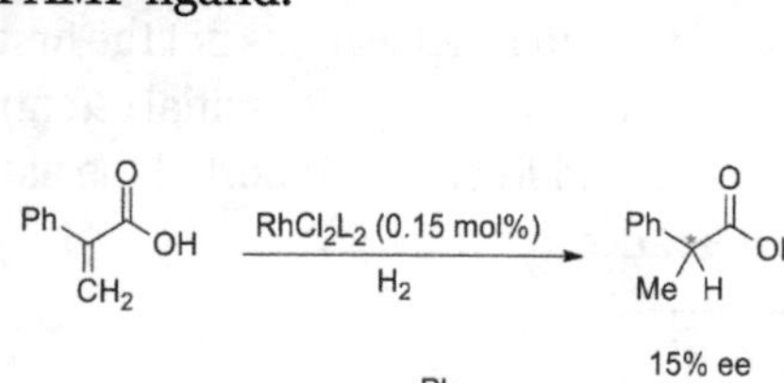

Knowles: Asymmetric hydrogenation (1968).

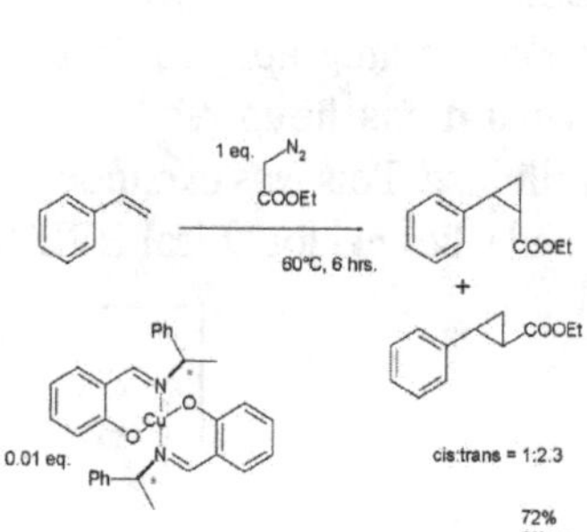

Noyori: Enantioselective cyclopropanation (1968).

Noyori devised a copper complex using a chiral Schiff base ligand, which he used for the metal-carbenoid cyclopropanation of styrene. In common with Knowles' findings, Noyori's results for the enantiomeric excess for this first-generation ligand were disappointingly low: 6%. However continued research eventually led to the development of the Noyori asymmetric hydrogenation reaction.

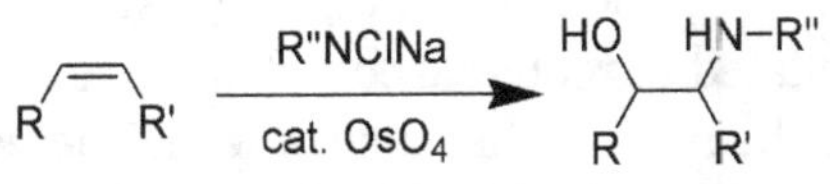

The Sharpless oxyamination.

Sharpless complemented these reduction reactions by developing a range of asymmetric oxidations (Sharpless epoxidation, Sharpless asymmetric dihydroxylation, Sharpless oxyamination) during the 1970s to 1980's. With the asymmetric oxyamination reaction, using osmium tetroxide, being the earliest.

During the same period, methods were developed to allow the analysis of chiral compounds by NMR; either using chiral derivatizing agents, such as Mosher's acid, or europium based shift reagents, of which $Eu(DPM)_3$ was the earliest.

Chiral auxiliaries were introduced by E.J. Corey in 1978 and featured prominently in the work of Dieter Enders. Around the same time enantioselective organocatalysis was developed, with pioneering work including the Hajos–Parrish–Eder–Sauer–Wiechert reaction. Enzyme-catalyzed enantioselective reactions became more and more common during the 1980s, particularly in industry, with their applications including asymmetric ester hydrolysis with pig-liver esterase. The emerging technology of genetic engineering has allowed the tailoring of enzymes to specific processes, permitting an increased range of selective transformations. For example, in the asymmetric hydrogenation of statin precursors.

Electrosynthesis

Electrosynthesis in chemistry is the synthesis of chemical compounds in an electrochemical cell. The main advantage of electrosynthesis over an ordinary redox reaction is avoidance of the potential wasteful other half-reaction and the ability to precisely tune the required potential. Electrosynthesis is actively studied as a science and also has many industrial applications. Electrooxidation is studied not only for synthesis but also for efficient removal of certain harmful organic compounds in wastewater.

Experimental Setup

The basic setup in electrosynthesis is a galvanic cell, a potentiostat and two electrodes. Good electrosynthetic conditions use a solvent and electrolyte combination that minimizes electrical resistance. Protic conditions often use alcohol-water or dioxane-water solvent mixtures with an electrolyte such as a soluble salt, acid or base. Aprotic conditions often use an organic solvent such as acetonitrile or dichloromethane with electrolytes such as lithium perchlorate or tetrabutylammonium acetate. Electrodes are selected which provide favorable electron transfer properties towards the substrate while maximizing the activation energy for side reactions. This activation energy is often related to an overpotential of a competing reaction. For example, in aqueous conditions the competing reactions in the cell are the formation of oxygen at the anode and hydrogen at the cathode. In this case a graphite anode and lead cathode could be used effectively because of their high overpotentials for oxygen and hydrogen formation respectively. Many other materials can be used as electrodes. Other examples include platinum, magnesium, mercury (as a liquid pool in the reactor), stainless steel or reticulated vitreous carbon. Some reactions use a sacrificial electrode is used which is consumed during the reaction like zinc or lead. The two basic cell types are undivided cell or divided cell type. In divided cells the cathode and anode chambers are separated with a semiporous membrane. Common membrane materials include sintered glass, porous porcelain, polytetrafluoroethene or polypropylene. The purpose of the divided cell is to permit the diffusion of ions while restricting the flow of the products and reactants. This is important when unwanted side reactions are possible. An example of a reaction requiring a divided cell is the reduction of nitrobenzene to phenylhydroxylamine, where the latter chemical is susceptible to oxidation at the anode.

Reactions

Organic oxidations take place at the anode with initial formation of radical cations as reactive intermediates. Compounds are reduced at the cathode to radical anions. The initial reaction takes place at the surface of the electrode and then the intermediates diffuse into the solution where they participate in secondary reactions.

The yield of an electrosynthesis is expressed both in terms the chemical yield and current efficiency. Current efficiency is the ratio of Coulombs consumed in forming the products to the total number of Coulombs passed through the cell. Side reactions decrease the current efficiency.

The potential drop between the electrodes determines the rate constant of the reaction. Electrosynthesis is carried out with either constant potential or constant current. The reason one chooses one over the other is due to a trade off of ease of experimental conditions versus current efficiency. Constant potential uses current more efficiently because the current in the cell decreases with time due to the depletion of the substrate around the working electrode (stirring is usually necessary to decrease the diffusion layer around the electrode). This is not the case under constant current conditions however. Instead as the substrate's concentration decreases the potential across the cell increases in order to maintain the fixed reaction rate. This consumes current in side reactions produced outside the target voltage.

Anodic Oxidations

- The most well-known electrosynthesis is the Kolbe electrolysis, in which two carboxylic acids decarboxylate, and the remaining structures bond together:

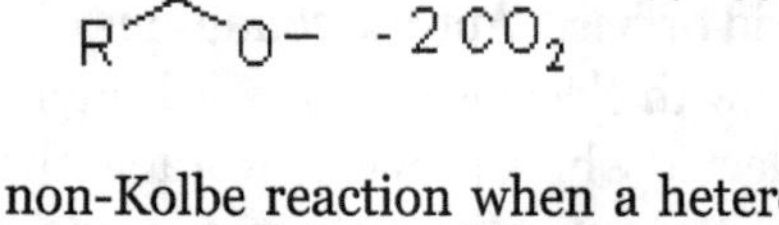

- A variation is called the non-Kolbe reaction when a heteroatom (nitrogen or oxygen) is present at the α-position. The intermediate oxonium ion is trapped by a nucleophile usually solvent.

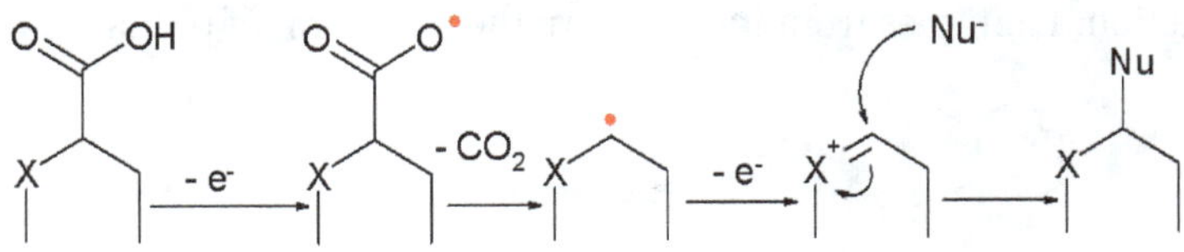

- Amides can be oxidized to *N*-acyliminium ions, which can be captured by various nucleophiles, for example:

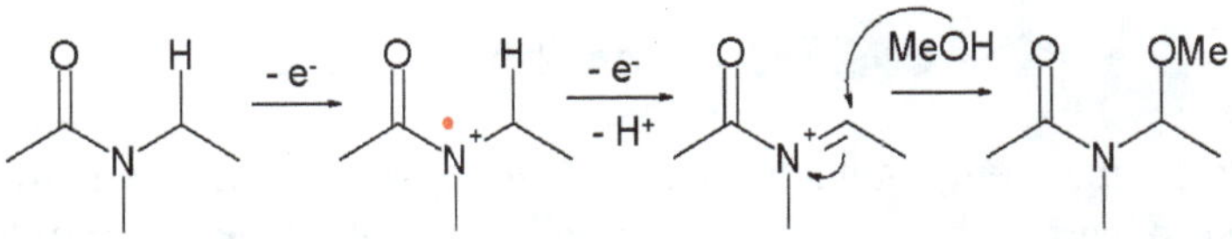

This reaction type is called a Shono oxidation. An example is the α-methoxylation of *N*-carbomethoxypyrrolidine:

- Oxidation of a carbanion can lead to a coupling reaction for instance in the electrosynthesis of the tetramethyl ester of ethanetetracarboxylic acid from the corresponding malonate ester
- α-amino acids form nitriles and carbon dioxide via oxidative decarboxylation at AgO anodes (the latter is formed *in-situ* by oxidation of Ag_2O):

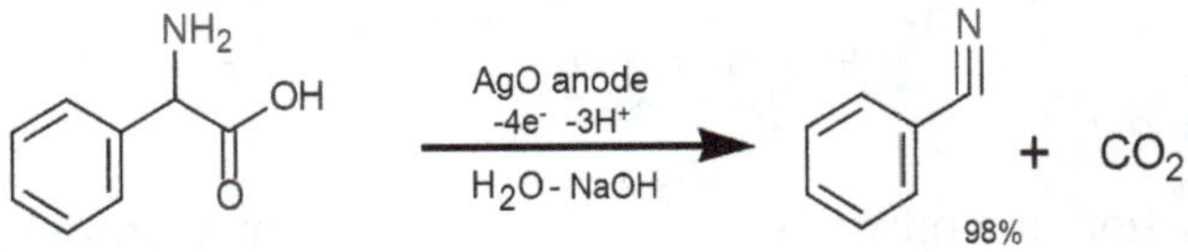

- Cyanoacetic acid from cathodic reduction of carbon dioxide and anodic oxidation of acetonitrile.

Cathodic Reductions

- In the Markó–Lam deoxygenation, an alcohol could be almost instantaneously deoxygenated by electroreducing their toluate ester.

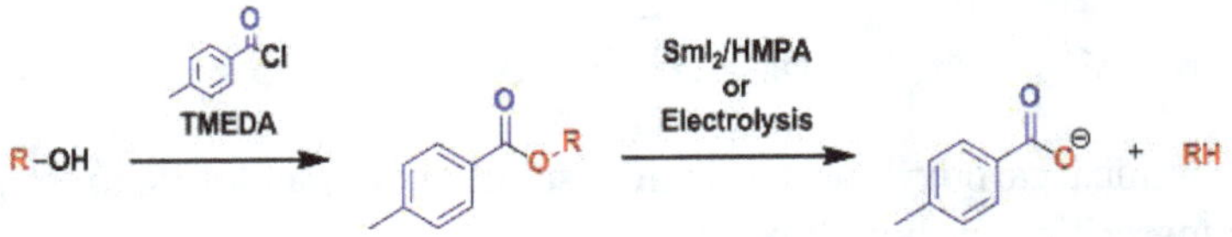

- The cathodic hydroisomerization of activated olefins is applied industrially in the synthesis of adiponitrile from 2 equivalents of acrylonitrile:

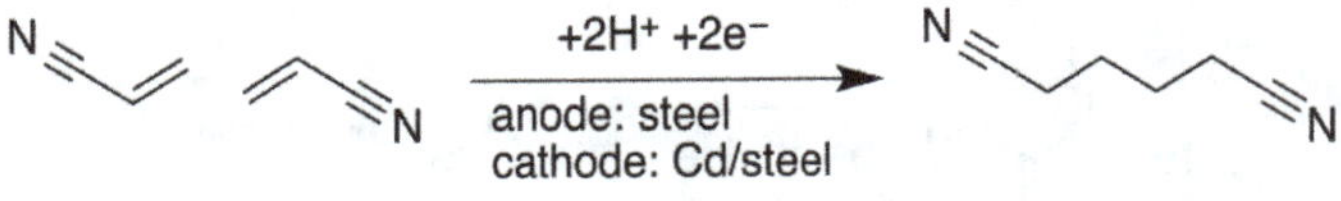

- The cathodic reduction of arene compounds to the 1,4-dihydro derivatives is similar to a

Birch reduction. Examples from industry are the reduction of phthalic acid:

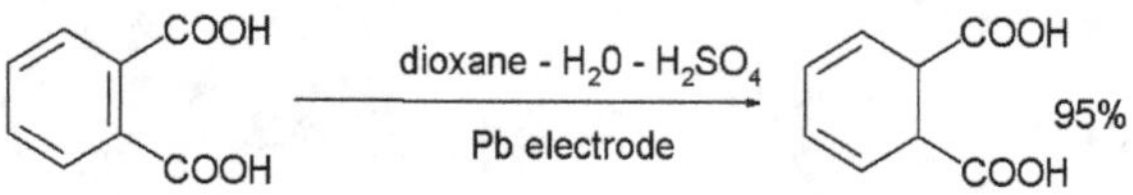

and the reduction of 2-methoxynaphthalene:

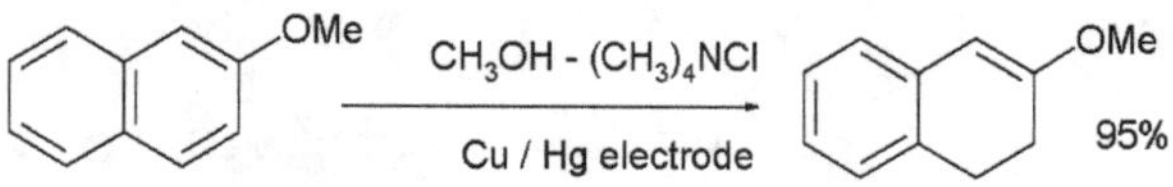

- The Tafel rearrangement, named for Julius Tafel, was at one time an important method for the synthesis of certain hydrocarbons from alkylated ethyl acetoacetate, a reaction accompanied by the rearrangement reaction of the alkyl group:

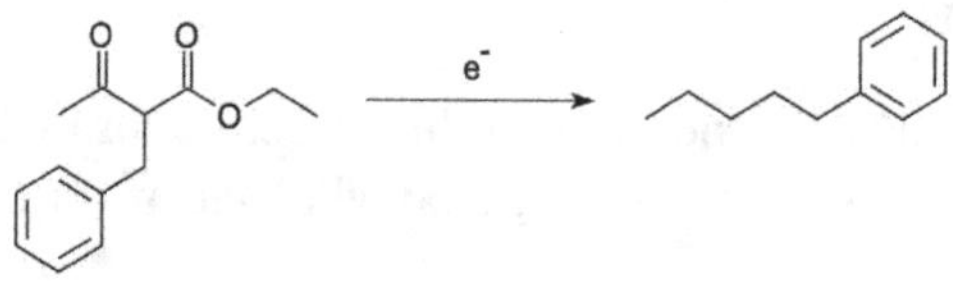

- The cathodic reduction of a nitrile to a primary amine in a divided cell:

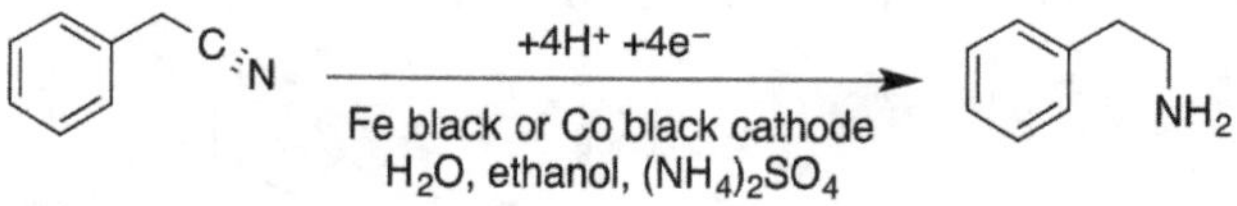

- Cathodic reduction of a nitroalkene can give the oxime in good yield. At higher negative reduction potentials, the nitroalkene can be reduced further, giving the primary amine but with lower yield.

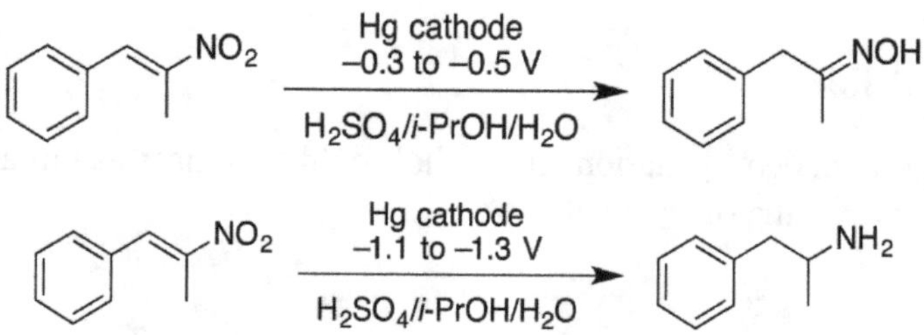

- An electrochemical carboxylation of a para-isobutylbenzyl chloride to Ibuprofen is promoted under supercritical carbon dioxide.

- Cathodic reduction of a carboxylic acid (oxalic acid) to an aldehyde (glyoxylic acid, shows as the rare aldehyde form) in a divided cell:

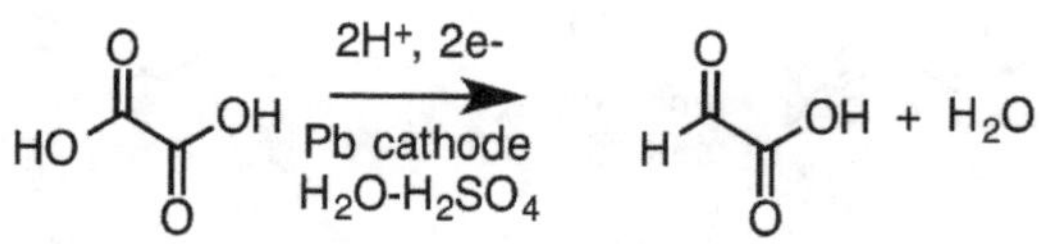

- An electrocatalysis by a copper complex helps reduce carbon dioxide to oxalic acid; this conversion uses carbon dioxide as a feedstock to generate oxalic acid.

Electrofluorination

In organofluorine chemistry, many perfluorinated compounds are prepared by electrochemical synthesis, which is conducted in liquid HF at voltages near 5–6 V using Ni anodes. The method was invented in the 1930s. Amines, alcohols, carboxylic acids, and sulfonic acids are converted to the perfluorinated derivatives using this technology. A solution or suspension of the hydrocarbon in hydrogen fluoride is electrolyzed at 5–6 V to produce high yields of the perfluorinated product.

References

- Madsen U, Krogsgaard-Larsen P, Liljefors T (2002). Textbook of Drug Design and Discovery. Washington, DC: Taylor & Francis. ISBN 0-415-28288-8
- Gröger, Albrecht Berkessel; Harald (2005). Asymmetric organocatalysis – from biomimetic concepts to applications in asymmetric synthesis (1. ed., 2. reprint. ed.). Weinheim: Wiley-VCH. ISBN 3-527-30517-3
- Wu-Pong S, Rojanasakul Y (2008). Biopharmaceutical drug design and development (2nd ed.). Totowa, NJ Humana Press: Humana Press. ISBN 978-1-59745-532-9
- Evans, D. A.; Helmchen, G.; Rüping, M. (2007). "Chiral Auxiliaries in Asymmetric Synthesis". In Christmann, M. Asymmetric Synthesis – The Essentials. Wiley-VCH Verlag GmbH & Co. pp. 3–9. ISBN 978-3-527-31399-0
- Koskinen, Ari M.P. (2013). Asymmetric synthesis of natural products (Second ed.). Hoboken, N.J.: Wiley. pp. 17, 28–29. ISBN 1118347331
- Reynolds CH, Merz KM, Ringe D, eds. (2010). Drug Design: Structure- and Ligand-Based Approaches (1 ed.). Cambridge, UK: Cambridge University Press. ISBN 978-0521887236
- Clayden, Jonathan; Greeves, Nick; Warren, Stuart; Wothers, Peter (2001). Organic Chemistry (1st ed.). Oxford University Press. ISBN 978-0-19-850346-0.Page 1226
- Grimshaw, James (2000). Electrochemical Reactions and Mechanisms in Organic Chemistry. Amsterdam: Elsevier Science. pp. 1–7, 282, & 310. ISBN 9780444720078
- Faber, Kurt (2011). Biotransformations in organic chemistry a textbook (6th rev. and corr. ed.). Berlin: Springer-Verlag. ISBN 9783642173936

5

Diverse Aspects of Organic Chemistry

Aromaticity, Covalent Bonding, Functional Groups and IUPAC Nomenclature are some of the basic concepts of organic chemistry. Aromaticity is the property of cycloalkenes where stabilization of the molecule is enhanced due to the ability of the electrons in the pi orbitals to delocalize. The topics elaborated in this chapter will help in gaining a better perspective about concepts of organic chemistry.

Aromaticity

Aromaticity is a property of conjugated cycloalkenes in which the stabilization of the molecule is enhanced due to the ability of the electrons in the π orbitals to delocalize. This act as framework to create a planar molecule.

Without aromatic compounds, we would not only be lacking many material necessities, our bodies would also not be able to function. Aromatic compounds are essential in industry; about 35 million tons of aromatic compounds are produced in the world every year to produce important chemicals and polymers, such as polyester and nylon. Aromatic compounds are also vital to the biochemistry of all living things. Three of the twenty amino acids used to form proteins ("the building blocks of life") are aromatic compounds and all five of the nucleotides that make up DNA and RNA sequences are all aromatic compounds. Needless to say, aromatic compounds are vital to us in many aspects.

The three general requirements for a compound to be aromatic are:

- The compound must be cyclic;
- Each element within the ring must have a p-orbital that is *perpendicular* to the ring, hence the molecule is planar;
- The compound must follow Hückel's Rule (the ring has to contain 4n+2 p-orbital electrons).

Among the many distinctive features of benzene, its aromaticity is the major contributor to why it is so unreactive. This topic will try to clarify the theory of aromaticity and why aromaticity gives unique qualities that make these conjugated alkenes inert to compounds such as Br_2 and even hydrochloric acid. It will also go into detail about the unusually large resonance energy due to the six conjugated carbons of benzene.

The delocalization of the p-orbital carbons on the sp^2 hybridized carbons is what gives the aromatic qualities of benzene.

Cyclic

Cycle or cyclic implies a ring. A cyclic molecule is one in which the first and last carbon in the chain are bound to each other.

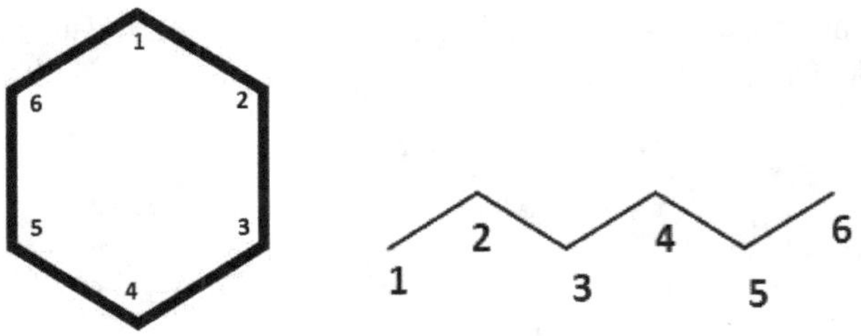

cyclohexane hexane

Carbon-1 is attached to carbon-2 but also attached to C-6; In hexane C-1 and C-6 are not attached to each other making it a linear compound.

Planar

Since aromaticity relies on the ability of orbitals to overlap, the ring must have its atoms in the same plane.

In short, the molecule must be flat.

Huckel's Rule: 4n+2 = Number of Resonating Electrons

$$4n + 2 = \text{Number of electrons}\,(e^-)$$

Resonating electrons include both pi electrons and lone pairs.

When solving for 'n', n must equal to a whole number. If we get a fraction then the molecule does not obey Huckel's rule.

Benzene is the most common aromatic molecule. A benzene ring has 3 pi bonds thus 6 resonating electrons, 4n+2 = 6

Solve for n:

$$4n + 2 = 6$$

$$4n = 4$$

$$n = 1$$

n = 1 which is a whole number.Benzene follows Huckel's rule.

Evidence of Aromaticity: Heats of Hydrogenation

One of the ways to test the relative amounts of resonance energy in a molecule is to compare the heats of hydrogenation between similar compounds. For instance, if we compare cyclohexene, 1,3-cyclohexadiene, and benzene, we would expect that their heats of hydrogenation will increase since the number of double bonds increases respectively. However, experimental evidence suggests that the actual heat of hydrogenation for benzene is actually 49.3 kcal/mole, making it even more stable than the 1,3-cyclohexadiene even though it has two double bonds, compared to benzene's three double bonds. This characteristic can be attributed to the aromaticity of benzene which delocalizes the electrons of the six pi orbitals.

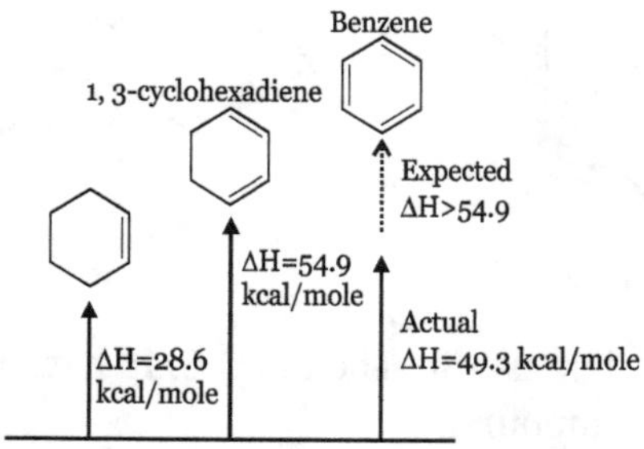

Aromatic compounds, originally named because of their fragrant properties, are unsaturated hydrocarbon ring structures that exhibit special properties, including unusual stability, due to their aromaticity. They are often represented as resonance structures containing single and double bonds. However, the bonding is stronger than expected for a conjugated structure, and it is more accurately depicted as delocalized electron density shared between all the atoms in the ring.

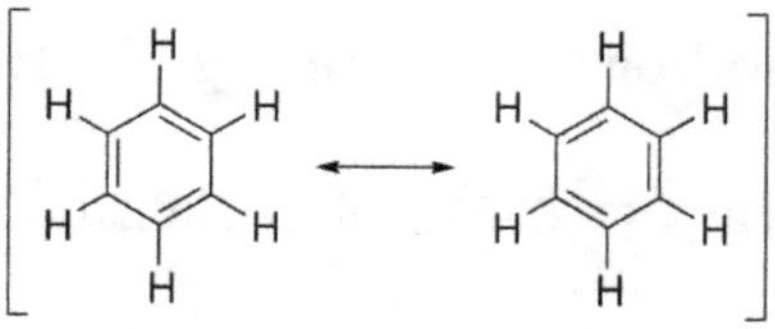

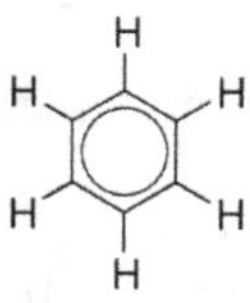

Benzene can only be fully depicted with all of its resonance structures, which show how its pi-electrons are delocalized throughout its six-carbon ring. This delocalization leads to a lower overall energy for the molecule, giving it greater stability.

Benzene

Benzene is symmetrical, planar, hexagonal strcture that boils at 80 C and melts at 5 C.

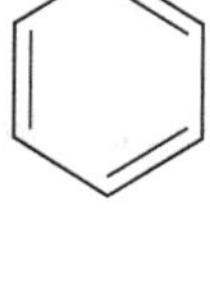

Benzene is the simplest of the aromatics (n=1 for Hückel's rule). Although you may see benzene written like that, benzene actually exists as two different compounds.

Kekulé structures for benzene

The benzene ring is constantly changing its form because of resonance. You can see the two structures that are possible for benzene.

Benzene Substituents

Two substituents that involve the benzene ring are phenyl- and benzyl- groups.

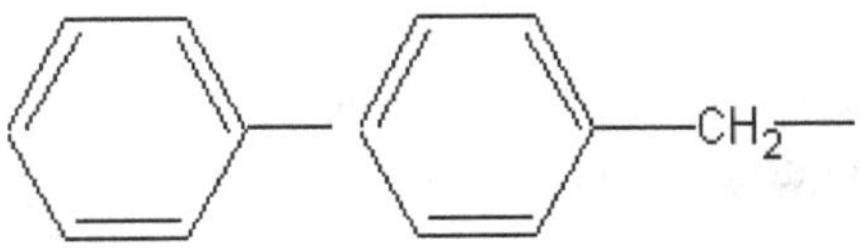

The molecule on the left is the phenyl- group and the one on the right is the benzyl- group. Aromatic groups are sometimes written Ar- (which is short for aryl) while R- represents an alkyl group. So if you saw someting written as Ar-CH_3, that would be a methylbenzene (or phenylmethane).

Importance of Aromatic Compounds

Aromatic compounds play key roles in the biochemistry of all living things. The four aromatic amino acids histidine, phenylalanine, tryptophan, and tyrosine each serve as one of the 20 basic building-blocks of proteins. Further, all 5 nucleotides (adenine, thymine, cytosine, guanine, and uracil) that make up the sequence of the genetic code in DNA and RNA are aromatic purines or pyrimidines. The molecule heme contains an aromatic system with 22 π-electrons. Chlorophyll also has a similar aromatic system.

Aromatic compounds are important in industry. Key aromatic hydrocarbons of commercial interest are benzene, toluene, *ortho*-xylene and *para*-xylene. About 35 million tonnes are produced worldwide every year. They are extracted from complex mixtures obtained by the refining of oil or by distillation of coal tar, and are used to produce a range of important chemicals and polymers, including styrene, phenol, aniline, polyester and nylon.

Types of Aromatic Compounds

The overwhelming majority of aromatic compounds are compounds of carbon, but they need not be hydrocarbons.

Neutral Homocyclics

Benzene, as well as most other annulenes (cyclodecapentaene excepted) with the formula C_nH_n where n is an even number, such as cyclotetradecaheptaene.

Heterocyclics

In heterocyclic aromatics (heteroaromatics), one or more of the atoms in the aromatic ring is of an element other than carbon. This can lessen the ring's aromaticity, and thus (as in the case of furan) increase its reactivity. Other examples include pyridine, pyrazine, pyrrole, imidazole, pyrazole, oxazole, thiophene, and their benzannulated analogs. In all these examples, the number of π-electrons is 6, due to the π-electrons from the double bonds as well as the two electrons from any lone pair that is in the p-orbital that is in the plane of the aromatic π system. For example, in pyridine, the five sp^2-hybridized carbons each have a p-orbital that is perpendicular to the plane of the ring, and each of these p-orbitals contains one π-electron. Additionally, the nitrogen atom is also sp^2-hybridized and has one electron in a p-orbital, which adds up to 6 p-electrons, thus making pyridine aromatic. The lone pair on the nitrogen is not part of the aromatic π system. Pyrrole and imidazole are both five membered aromatic rings that contain heteroatoms. In pyrrole, each of the four sp^2-hybridized carbons contributes one π-electron, and the nitrogen atom is also sp^2-hybridized and contributes two π-electrons from its lone pair, which occupies a p-orbital. In imidazole, both nitrogens are sp^2-hybridized; the one in the double bond contributes one electron and the one which is not in the double bond and is in a lone pair contributes two electrons to the π system.

Fused Aromatics and Polycyclics

Polycyclic aromatic hydrocarbons are molecules containing two or more simple aromatic rings fused together by sharing two neighboring carbon atoms. Examples are naphthalene, anthracene, and phenanthrene. In fused aromatics, not all carbon–carbon bonds are necessarily equivalent, as the electrons are not delocalized over the entire molecule. The aromaticity of these molecules can be explained using their orbital picture. Like benzene and other monocyclic aromatic molecules, polycyclics have a cyclic conjugated pi system with p-orbital overlap above and below the plane of the ring.

Substituted Aromatics

Many chemical compounds are aromatic rings with other functional groups attached. Examples include trinitrotoluene (TNT), acetylsalicylic acid (aspirin), paracetamol, and the nucleotides of DNA.

Aromatic Ions

Aromatic molecules need not be neutral molecules. Ions that satisfy Huckel's rule of $4n + 2$ π-electrons in a planar, cyclic, conjugated molecule are considered to be aromatic ions. For example, the cyclopentadienyl anion and the cycloheptatrienylium cation are both considered to be aromatic ions, and the azulene molecule can be approximated as a combination of both.

In order to convert the atom from sp^3 to sp^2, a carbocation, carbanion, or carbon radical must be formed. These leave sp^2-hybridized carbons that can partake in the π system of an aromatic molecule. Like neutral aromatic compounds, these compounds are stable and form easily. The cyclopentadienyl anion is formed very easily and thus 1, 3-cyclopentadiene is a very acidic hydrocarbon with a pK_a of 16. Other examples of aromatic ions include the cyclopropenium cation (2 π-electrons) and cyclooctatetraenyl dianion (10 pi electrons).

Atypical Aromatic Compounds

Aromaticity also occurs in compounds that are not carbocyclic or heterocyclic; inorganic six-membered-ring compounds analogous to benzene have been synthesized. For example, borazine is a six-membered ring composed of alternating boron and nitrogen atoms, each with one hydrogen attached. It has a delocalized π system and undergoes electrophilic substitution reactions appropriate to aromatic rings rather than reactions expected of non-aromatic molecules.

Quite recently, the aromaticity of planar Si_5^{6-} rings occurring in the Zintl phase $Li_{12}Si_7$ was experimentally evinced by Li solid-state NMR. Metal aromaticity is believed to exist in certain clusters of aluminium.

Homoaromaticity is the term used to describe systems where conjugation is interrupted by a single sp^3 hybridized carbon atom.

Möbius aromaticity occurs when a cyclic system of molecular orbitals, formed from p_π atomic orbitals and populated in a closed shell by $4n$ (n is an integer) electrons, is given a single half-twist to form a Möbius strip. A π system with $4n$ electrons in a flat (non-twisted) ring would be antiaromatic, and therefore highly unstable, due to the symmetry of the combinations of p atomic orbitals. By twisting the ring, the symmetry of the system changes and becomes allowed, because the twist can be left-handed or right-handed, the resulting Möbius aromatics are *dissymmetric* or chiral. But as of 2012, no Möbius aromatic molecules had been synthesized. Aromatics with two half-twists corresponding to the paradromic topologies were first suggested by Johann Listing. In carbo-benzene the ring bonds are extended with alkyne and allene groups.

Y-aromaticity is a concept developed to explain the extraordinary stability and high basicity of the guanidinium cation. Guanidinium is not a ring molecule, and is cross-conjugated rather than a linear π system, but is reported to have its six π-electrons delocalized over the whole molecule. The concept is controversial and some authors emphasize different effects.

σ-aromaticity refers to stabilization arising from the delocalization of sigma bonds. It is often invoked in cluster chemistry and is closely related to Wade's Rule.

Covalent Bonding

In Chemistry, Covalent bond is the interatomic linkage that results from the sharing of an electron pair between two atoms.

The binding arises from the electrostatic attraction of their nuclei for the same electrons. A covalent bond forms when the bonded atoms have a lower total energy than that of widely separated atoms.

Molecules that have covalent linkages include the inorganic substances hydrogen, nitrogen, chlorine, water, and ammonia (H_2, N_2, Cl_2, H_2O, NH_3) together with all organic compounds. In structural representations of molecules, covalent bonds are indicated by solid lines connecting pairs of atoms:

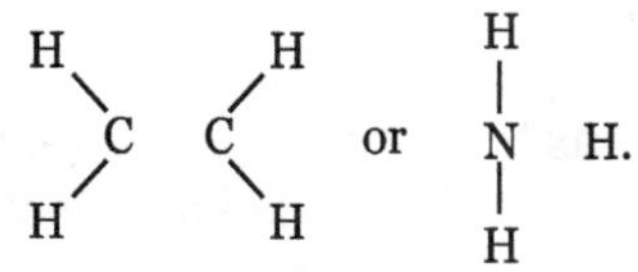

A single line indicates a bond between two atoms (*i.e.*, involving one electron pair), double lines (=) indicate a double bond between two atoms (*i.e.*, involving two electron pairs), and triple lines (≡) represent a triple bond, as found, for example, in carbon monoxide (C≡O). Single bonds consist of one sigma (σ) bond, double bonds have one σ and one pi (π) bond, and triple bonds have one σ and two π bonds.

The idea that two electrons can be shared between two atoms and serve as the link between them was first introduced in 1916 by the American chemist G.N. Lewis, who described the formation of such bonds as resulting from the tendencies of certain atoms to combine with one another in order for both to have the electronic structure of a corresponding noble-gas atom.

Covalent bonds are directional, meaning that atoms so bonded prefer specific orientations relative to one another; this in turn gives molecules definite shapes, as in the angular (bent) structure of the H_2O molecule. Covalent bonds between identical atoms (as in H_2) are nonpolar—*i.e.*, electrically uniform—while those between unlike atoms are polar—*i.e.*, one atom is slightly negatively charged and the other is slightly positively charged. This partial ionic character of covalent bonds increases with the difference in the electro negativities of the two atoms.

Covalent Structures

There are several types of structures for covalent substances, including individual molecules, molecular structures, macromolecular structures and giant covalent structures. Individual molecules have strong bonds that hold the atoms together, but there are negligible forces of attraction between molecules. Such covalent substances are usually gases, for example, HCl, SO_2, CO_2, and CH_4. In molecular structures, there are weak forces of attraction. Such covalent substances are low-boiling-temperature liquids (such as ethanol), and low-melting-temperature solids (such as iodine and solid CO_2). Macromolecular structures have large numbers of atoms linked by covalent bonds in chains, including synthetic polymers such as polyethylene and nylon, and biopolymers such as proteins and starch. Network covalent structures (or giant covalent structures) contain large numbers of atoms linked in sheets (such as graphite), or 3-dimensional structures (such as diamond and quartz). These substances have high melting and boiling points, are frequently brittle, and tend to have high electrical resistivity. Elements that have high electronegativity, and the ability to form three or four electron pair bonds, often form such large macromolecular structures.

One-and Three-electron Bonds

Bonds with one or three electrons can be found in radical species, which have an odd number of electrons. The simplest example of a 1-electron bond is found in the dihydrogen cation H_2^+ One-electron bonds often have about half the bond energy of a 2-electron bond, and are therefore called "half bonds". However, there are exceptions: in the case of dilithium, the bond is actually stronger for the 1-electron Li_2^+ than for the 2-electron Li_2. This exception can be explained in terms of hybridization and inner-shell effects.

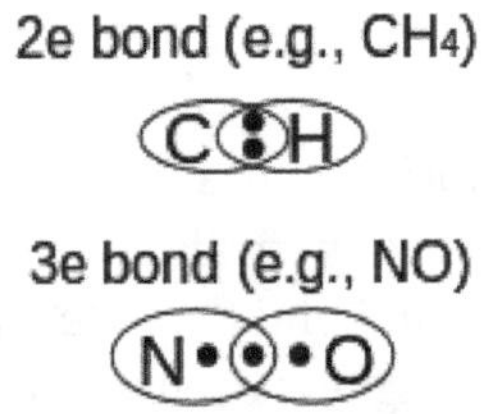

Comparison of the electronic structure of the three-electron bond to the conventional covalent bond.

The simplest example of three-electron bonding can be found in the helium dimer cation He_2^+. It is considered a "half bond" because it consists of only one shared electron (rather than two); in molecular orbital terms, the third electron is in an anti-bonding orbital which cancels out half of the bond formed by the other two electrons. Another example of a molecule containing a 3-electron bond, in addition to two 2-electron bonds, is nitric oxide, NO. The oxygen molecule, O_2 can also be regarded as having two 3-electron bonds and one 2-electron bond, which accounts for its paramagnetism and its formal bond order of 2. Chlorine dioxide and its heavier analogues bromine dioxide and iodine dioxide also contain three-electron bonds.

Molecules with odd-electron bonds are usually highly reactive. These types of bond are only stable between atoms with similar electro negativities.

Resonance

There are situations whereby a single Lewis structure is insufficient to explain the electron configuration in a molecule, hence a superposition of structures are needed. The same two atoms in such molecules can be bonded differently in different structures (a single bond in one, a double bond in another, or even none at all), resulting in a non-integer bond order. The nitrate ion is one such example with three equivalent structures. The bond between the nitrogen and each oxygen is a double bond in one structure and a single bond in the other two, so that the average bond order for each N–O interaction is: $(2+1+1)/3 = 4/3$.

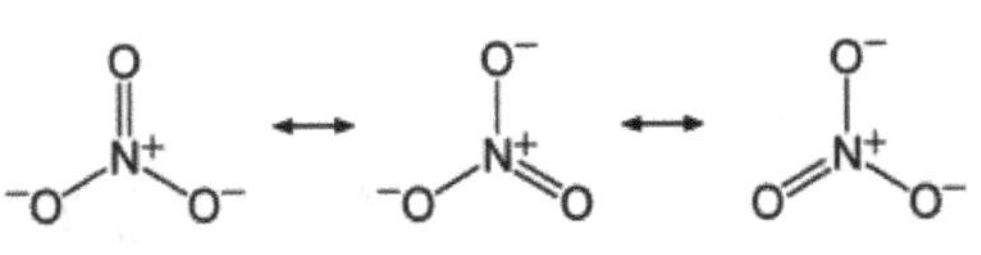

Aromaticity

In organic chemistry, when a molecule with a planar ring obeys Hückel's rule, where the number of π electrons fit the formula $4n + 2$ (where n is an integer), it attains extra stability and symmetry. In benzene, the prototypical aromatic compound, there are 6 π bonding electrons ($n = 1$, $4n + 2 = 6$). These occupy three delocalized π molecular orbitals (molecular orbital theory) or form conjugate π bonds in two resonance structures that linearly combine (valence bond theory), creating a regular hexagon exhibiting a greater stabilization than the hypothetical 1, 3, 5-cyclohexatriene.

In the case of heterocyclic aromatics and substituted benzenes, the electronegativity differences between different parts of the ring may dominate the chemical behavior of aromatic ring bonds, which otherwise are equivalent.

Hypervalence

Certain molecules such as xenon difluoride and sulfur hexafluoride have higher co-ordination numbers than would be possible due to strictly covalent bonding according to the octet rule. This is explained by the three-center four-electron bond ("3c–4e") model which interprets the molecular wave function in terms of non-bonding highest occupied molecular orbitals in molecular orbital theory and ionic-covalent resonance in valence bond theory.

Electron-deficiency

In three-center two-electron bonds ("3c–2e") three atoms share two electrons in bonding. This type of bonding occurs in electron deficient compounds like diborane. Each such bond (2 per molecule in diborane) contains a pair of electrons which connect the boron atoms to each other in a banana shape, with a proton (nucleus of a hydrogen atom) in the middle of the bond, sharing electrons with both boron atoms. In certain cluster compounds, so-called four-center two-electron bonds also have been postulated.

Quantum Mechanical Description

After the development of quantum mechanics, two basic theories were proposed to provide a quantum description of chemical bonding: valence bond (VB) theory and molecular orbital (MO) theory. A more recent quantum description is given in terms of atomic contributions to the electronic density of states.

Covalency from Atomic Contribution to the Electronic Density of States

In COOP, COHP and BCOOP, evaluation of bond covalency is dependent on the basis set. To overcome this issue, an alternative formulation of the bond covalency can be provided in this way.

The center mass $cm(n,l,m_l,m_s)$ of an atomic orbital $|n,l,m_l,m_s\rangle$, with quantum numbers n, l, m_l, m_s, for atom A is defined as:

$$cm^{\mathrm{A}}(n,l,m_l,m_s)=\frac{\int_{E_0}^{E_1} E g^{\mathrm{A}}_{|n,l,m_l,m_s\rangle}(E)dE}{\int_{E_0}^{E_1} g^{\mathrm{A}}_{|n,l,m_l,m_s\rangle}(E)dE}$$

Where, $g^{\mathrm{A}}_{|n,l,m_l,m_s\rangle}(E)$ E is the contribution of the atomic orbital $|n,l,m_l,m_s\rangle$ of the atom A to the total electronic density of states $g(E)$ of the solid:

$$g(E)=\sum_{\mathrm{A}}\sum_{n,l}\sum_{m_l,m_s} g^{\mathrm{A}}_{|n,l,m_l,m_s\rangle}(E)$$

Where, the outer sum runs over all atoms A of the unit cell. The energy window $[E_0, E_1]$ is chosen in such a way that it encompasses all relevant bands participating in the bond. If the range to select is unclear, it can be identified in practice by examining the molecular orbitals that describe the electron density along the considered bond.

The relative position $C_{n_A l_A, n_B l_B}$ of the center mass of $|n_A, l_A\rangle$ levels of atom A with respect to the center mass of $|n_B, l_B\rangle$ levels of atom B is given as:

$$C_{n_A l_A, n_B l_B} = -\left| cm^{A}(n_A, l_A) - cm^{B}(n_B, l_B) \right|$$

Where, the contributions of the magnetic and spin quantum numbers are summed. According to this definition, the relative position of the A levels with respect to the B levels is,

$$C_{A,B} = -\left| cm^{A} - cm^{B} \right|$$

Where, for simplicity, we may omit the dependence from the principal quantum number n in the notation referring to $C_{n_A l_A, n_B l_B}$

In this formalism, the greater the value of $C_{A,B}$, the higher the overlap of the selected atomic bands, and thus the electron density described by those orbitals gives a more covalent A–B bond. The quantity $C_{A,B}$ is denoted as the *covalency* of the A–B bond, which is specified in the same units of the energy E.

Single Bonds

A single bond is when two electrons one pair of electrons are shared between two atoms. It is depicted by a single line between the two atoms. Although this form of bond is weaker and has a smaller density than a double bond and a triple bond, it is the most stable because it has a lower level of reactivity meaning less vulnerability in losing electrons to atoms that want to steal electrons.

Example: HCL

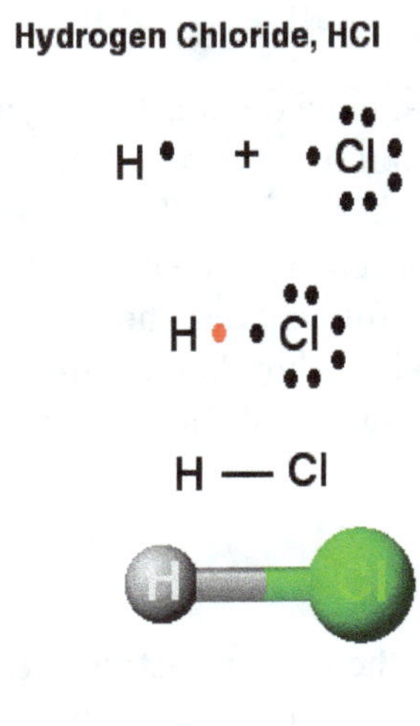

Double Bonds

A Double bond is when two atoms share two pairs of electrons with each other. It is depicted by two horizontal lines between two atoms in a molecule. This type of bond is much stronger than a single bond, but less stable; this is due to its greater amount of reactivity compared to a single bond.

Example: CO_2

Below is a Lewis dot structure of Carbon dioxide demonstrating a double bond. As you can see from the picture below, Carbon dioxide has a total of 1 Carbon atom and 2 Oxygen atoms. Each Oxygen atom has 6 valence electrons whereas the Carbon atom only has 4 valence electrons. To satisfy the Octet Rule, Carbon needs 4 more valence electrons. Since each Oxygen atom has 3 lone pairs of electrons, they can each share 1 pair of electrons with Carbon; as a result, filling Carbon's outer valence shell (Satisfying the Octet Rule).

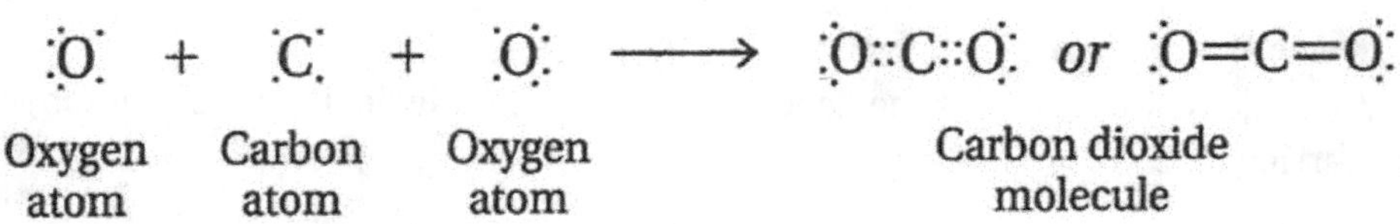

Triple Bond

A Triple bond is when three pairs of electrons are shared between two atoms in a molecule. It is the least stable out of the three general types of covalent bonds. It is very vulnerable to electron thieves.

Example: Acetylene

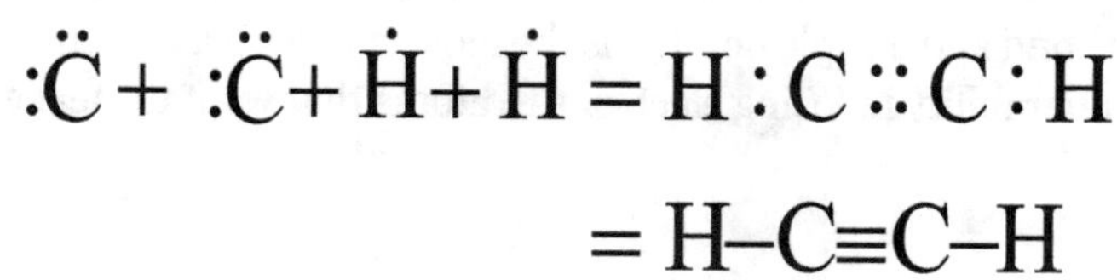

Below is a Lewis dot structure of Acetylene demonstrating a triple bond. As you can see from the picture below, Acetylene has a total of 2 Carbon atoms and 2 Hydrogen atoms. Each Hydrogen atom has 1 valence electron whereas each Carbon atom has 4 valence electrons. Each Carbon needs 4 more electrons and each Hydrogen needs 1 more electron. Hydrogen shares its only electron with Carbon to get a full valence shell. Now Carbon has 5 electrons. Because each Carbon atom has 5 electrons -1 single bond and 3 unpaired electrons the two Carbons can share their unpaired electrons, forming a triple bond. Now all the atoms are happy with their full outer valence shell.

Polar Covalent Bond

A Polar Covalent Bond is created when the shared electrons between atoms are not equally shared. This occurs when one atom has a higher electronegativity than the atom it is sharing with. The atom with the higher electronegativity will have a stronger pull for electrons (Similiar to a Tug-O-War game, whoever is stronger usually wins). As a result, the shared electrons will be closer to the atom with the higher electronegativity, making it unequally shared. A polar covalent bond will result in the molecule having a slightly positive side (the side containing the atom with a lower electronegativity) and a slightly negative side (containing the atom with the higher electronegativity) because the shared electrons will be displaced toward the atom with the higher electronegativity. As a result of polar covalent bonds, the covalent compound that forms will have an electrostatic potential. This potential will make the resulting molecule slightly polar, allowing it to form weak bonds with other polar molecules. One example of molecules forming weak bonds with each other as a result of an unbalanced electrostatic potential is hydrogen bonding, where a hydrogen atom will interact with an electronegative hydrogen, fluorine, or oxygen atom from another molecule or chemical group.

Example: Water, Sulfide, Ozone, etc.

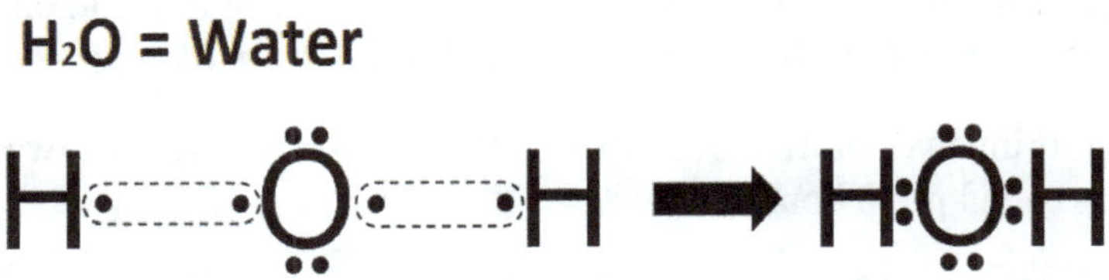

Water is a molecule which has the molecular formula of H2O. Each hydrogen atom bonds with a central oxygen atom by sharing a pair of electrons through a covalent bond. The result is a bent molecule with a mean H-O-H angle of 106°.

Nonpolar Covalent Bond

A Nonpolar Covalent Bond is created when atoms share their electrons equally. This usually occurs when two atoms have similar or the same electron affinity. The closer the values of their electron affinity, the stronger the attraction. This occurs in gas molecules; also known as diatomic elements. Nonpolar covalent bonds have a similar concept as polar covalent bonds; the atom with the higher electronegativity will draw away the electron from the weaker one. Since this statement is true if we apply this to our diatomic molecules all the atoms will have the same electronegativity since they are the same kind of element; thus, the electro negativities will cancel each other out and will have a charge of 0(A.K.A. Nonpolar covalent bond).

Examples of gas molecules that have a nonpolar covalent bond: Hydrogen gas atom, Nitrogen gas atoms, etc.

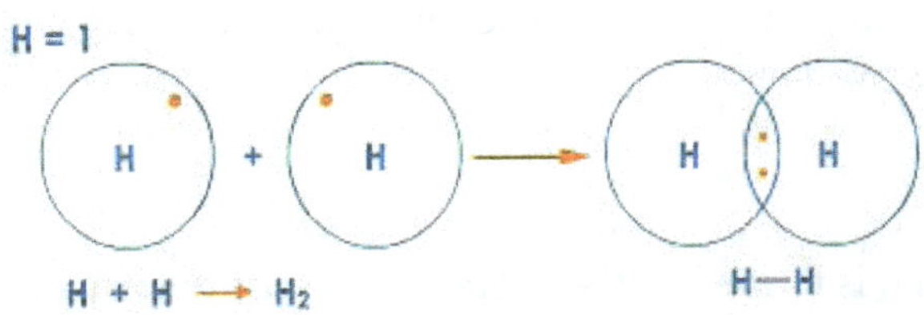

As you can see from the picture above, Hydrogen gas has a total of 2 Hydrogen atoms. Each Hydrogen atom has 1 valence electron. Since Hydrogen can only fit a max of 2 valence electrons in its orbital, each Hydrogen atom only needs 1 electron. Each atom has 1 valence electron, so they can just share, giving each atom two electrons each.

Sigma (σ) and Pi bonds

Sigma (σ) Bond

This type of covalent bond is formed by the axial overlapping of half-filled atomic orbitals. The atomic orbitals overlap along the inter-nuclear axis and involve end to end or head' on overlap. The electron cloud formed as a result of axial overlap is cylindrically symmetrical about inter-nuclear axis. The electrons constituting sigma bond are called sigma electrons. There can be three types of axial overlap among s and p-orbitals as discussed below:

- s-s overlap: It involves mutual overlap of half-filled s-orbitals of the atoms approaching to form a bond. The bond formed is called s-s σ bond.
- s-p overlap: It involves mutual overlap of half-filled s-orbital of the one atom with half-filled p-orbital of the other. The bond so formed is called s-p σ bond.
- p-p overlap: It involves mutual overlap of half-filled p-orbitals of the two atoms. The bond so formed is called p-p σ bond.

The s -s, s-p and p-p overlaps have been shown diagrammatically in Figure below:

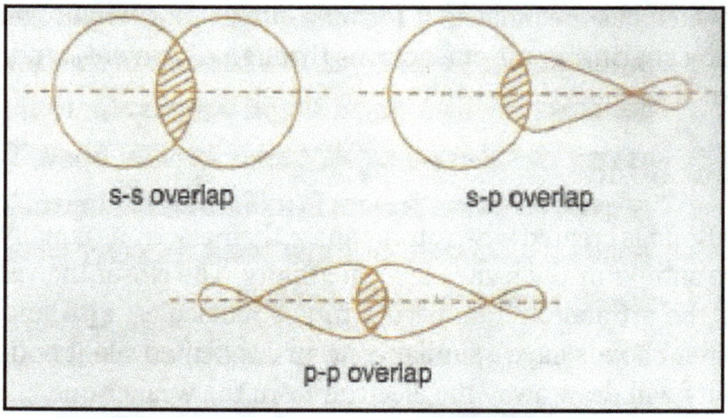

Figure: s-s, s-p and p-p overlap of atomic orbitals.

Strength of three types of sigma bonds, the strength of three types of sigma bonds varies as follows:

p-p > p-s > s-s

It is because of the fact, that p-orbitals allow overlap to greater extent as compared top-s which is larger as compared to s-s overlap.

Pi Bond

This type of covalent bond is formed by the lateral or sidewise overlap of the atomic orbitals. The orbital overlap takes place in such a way that their axes are parallel to each other but perpendicular to the inter-nuclear axis. The pi bond consists of two charge clouds above and below the plane of the atoms involved in the bond formation. The electrons involved in the n-bond formation are called n-electrons.

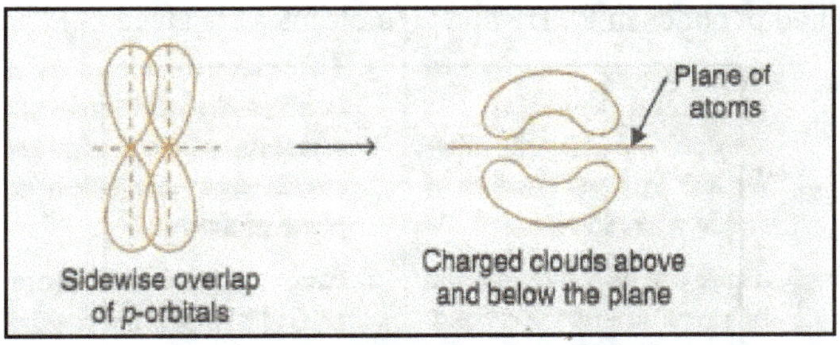

Figure: Sidewise overlap of p-orbitals.

It may be noted that:

- Sigma bond is stronger than pi bond, It is because of the fact that overlapping of atomic orbitals can take place to a greater extent during the formation of sigma bond whereas overlapping of orbitals occurs to a smaller extent during the formation of pi bond.
- Pi bond between the two atoms is formed only in addition to a sigma bond. It is because of the fact that the atoms constituting a single bond prefer to form a strong sigma bond rather than a weak pi bond. Thus, pi bond is always present in molecules having multiple bonds, i.e., double or triple bond. In other words, a single bond cannot be a pi bond.
- The shape of molecule is controlled by the sigma framework (orientations of sigma bonds) around the central atom. Pi bonds are superimposed on sigma bonds hence they simply modify the dimensions of the molecule.

Comparison Between Sigma and Pi Bonds

The various points of distinction between sigma and pi bonds are given in table.

Sigma. Bond	Pi Bond
1. This bond is formed by the axial overlap of atomic orbitals.	1. The bond is formed by the sidewise overlap of atomic orbitals.
2. This bond can be formed by overlap of s-s, s-p or p-p orbitals.	2. It involves of overlap of p-orbitals only.
3. The bond is stronger because overlapping can take place to a larger extent.	3. The bond is weaker because the overlapping occurs to a smaller extent.
4. The electron cloud formed by axial overlap is symmetrical about the inter-nuclear axis and consists of single charged cloud.	4. The electron cloud of pi bond is discontinuous and consists of two charged clouds above and below the plane of atoms.
5. There can be a free rotation of atoms around the σ bond.	5. Free rotation of atoms around π bond is not possible because it involves breaking of π bond.
6. The bond may be present between the two atoms either alone or along with π bond.	6. The bond is always present between the two atoms along with the sigma bond, i.e., it is always superimposed on sigma bond.
7. The shape of molecule is determined by the sigma framework around the central atom.	7. The π bonds do not contribute to the shape, however, they modify the dimensions of the molecule.

Formation of sigma and pi bonds in F_2, HF and O_2 molecules are given in Figure below:

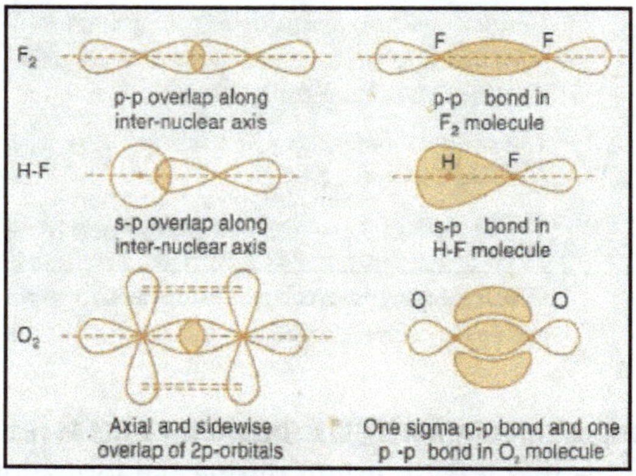

Fig: Formation of F_2, HF and o_2 molecules.

Coordinate Covalent Bond

A coordinate bond is a type of covalent bond where both of the electrons that form the bond originate from the same atom (more generally, a "dative" covalent bond).

Coordinate bonds form between a central electrophile (low electron density, such as metal cations) and one or more nucleophiles (high electron density, such as the hydroxide anion) oriented around the former. Nucleophiles acts as "ligands" by supplying two electrons per coordinate bond to the electrophile to satisfy the octet rule. A "polydentate" ligand molecule may form multiple coordinate bonds.

As with any covalent bond, the atoms in a coordinate bond redistribute electrons in order to satisfy the octet rule, which states that each atom will lose, gain, or share electrons in order to have a full valence of eight electrons in its outer shell. Hydrogen is a common exception to the octet rule, and instead follows the duet rule because it only has one 1s orbital and its outer valence is full when it has two electrons.

It is important to note that coordinate bonding between oppositely charged ionic species is of a covalent character, where the bonding electrons are shared, as opposed to ionic compounds.

Coordinate Covalent Bond (Dative Bond) Formation

One of the two atoms must have at least one lone pair of electrons, e.g., ammonia (NH_3), water(H_2O). A pair of electrons which is not shared with any other atom is known as the lone pair of electrons but it is provided to the other atom for the formation of the coordinate Covalent bond.

Another atom should be short of at least a lone pair of electrons, e.g., Hydrogen ion (H^+).

Examples

Ammonium Ion Coordinate Bond (NH_3)

The Nitrogen atom has five valence electrons and hydrogen atom has one valence electron. In ammonia, each of the three hydrogen atoms is bonded to the nitrogen atom by a pair of shared electrons. (One electron from the hydrogen atom and one from the nitrogen).

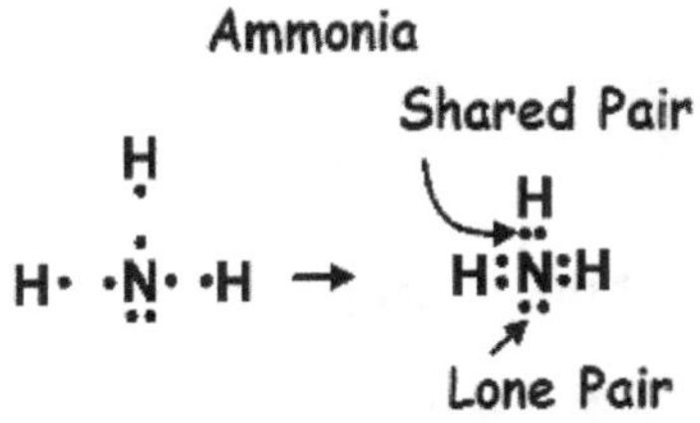

Thus, the nitrogen atom in ammonia is left alone pair of electrons.

This ammonia molecule combines with a hydrogen ion H^+, which has no electrons in its outermost orbit in such a way that the lone pair of electrons on the nitrogen atom is shared by the hydrogen ion. The latter thus acquires the helium configuration. This bond which is formed between the nitrogen atom in ammonia and the hydrogen ion is a coordinate bond.

Lewis Structure of Ammonium Ion($NH4^+$)

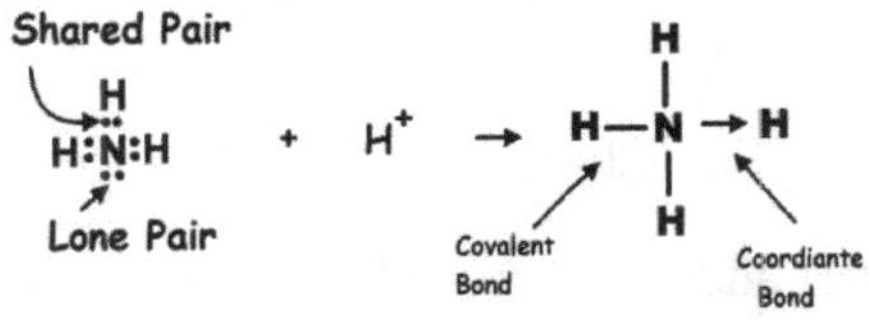

The nitrogen atom in an ammonia molecule contains a lone pair of electrons. It is a donor. The hydrogen ion (H^+) accepts that lone pair of electrons (Acceptor) and forms a coordinate bond. Once this bond is formed, all four N-H bonds become identical. The linkage forms an ammonium having a single positive charge.

When ammonium chloride NH_4Cl is formed cation $NH4^+$ (having 3 covalent and one coordinate bond) and anion Cl^- are attracted towards each other due to electrical charge existing between them ionic bond is formed. Thus ammonium chloride is a good example of a compound having all the three types of bonds i.e., covalent, coordinate and ionic bond.

Hydronium Ion Coordinate Covalent Bond(H_3O^+)

Water contains two hydrogen atoms and one oxygen atom. The hydrogen atom has one electron in its valence shell while the oxygen atom has six electrons in its Valence shell.

Lewis Structure of H_2O

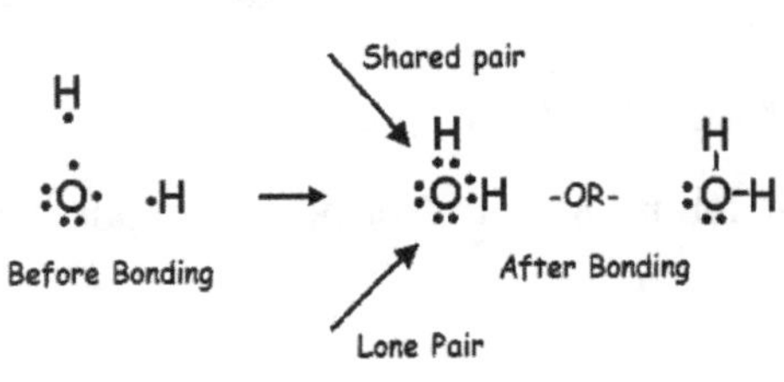

Two hydrogen atoms share one electron each with the oxygen atom to form a water molecule. The two unshared pairs of electrons known as lone pairs of electrons, do not participate in the bond formation in the water molecule.

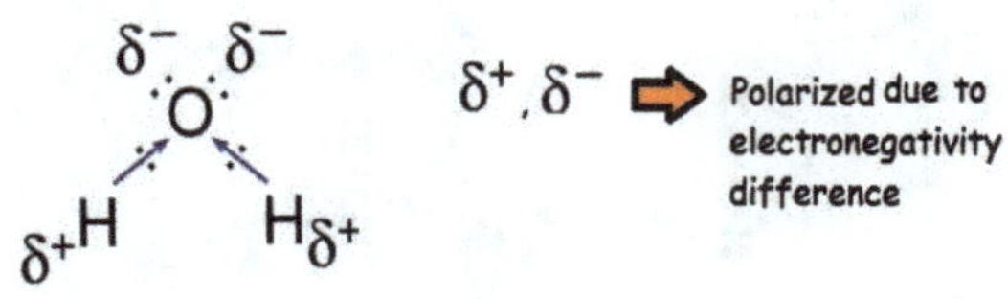

The O-H bonds in a water molecule are polarised due to the large difference in the electronegativity values of the oxygen and the hydrogen atoms. (The oxygen atom has an electronegativity value of 3.5, while that of the hydrogen atom is 2.1). Oxygen has higher electronegativity so it attained slightly negative charge while hydrogen atoms attain slight positive charges. Therefore, a molecule of water due to these charges is a polar molecule and is represented as shown alongside.

When an acid is added to water, strong dipole interaction (attraction between slightly positive and slight negative charge) occurs between the polar water molecules and the polar acid molecules.

As a result of such a dipole interaction, the hydrogen ion gets released from the polar bonds of the acid molecules:

$$HCl + H_2O \rightarrow H^+ + Cl^-$$

The hydrogen ion spontaneously adds on to one of the two lone pairs of electrons of the oxygen atom of the water molecule.

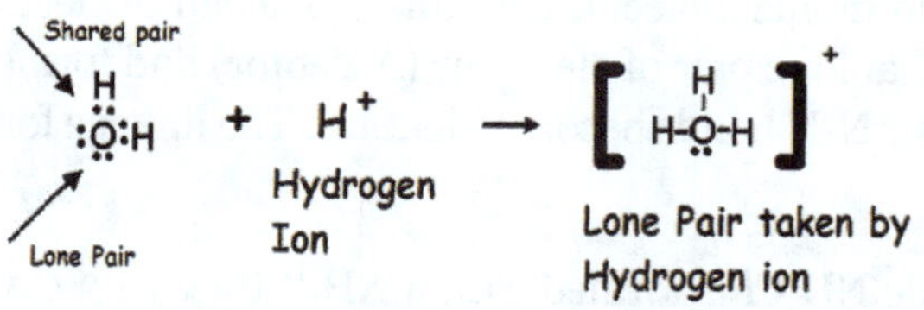

An addition of released H^+ ion to a lone pair of electrons of the oxygen atom of the polar water molecule leads to the formation of a hydronium ion.

Proton + Water → Hydronium ion

$H^+ + H_2O \rightarrow H_3O^+$

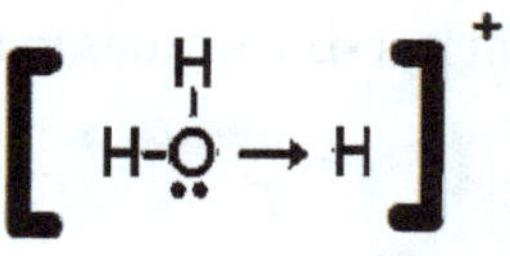

Hydroxyl ion coordinate covalent bond OH^-.

The hydroxyl ion or hydroxide ion is OH^-. It is formed when one hydrogen ion (H^+) is removed from the water molecule.

$H_2O \rightarrow H^+ + OH^-$

When H^+ is removed from the water molecule, the shared pair of electron remains with oxygen.

As, oxygen is more electronegative and thus hydroxyl ion has the negative charge.

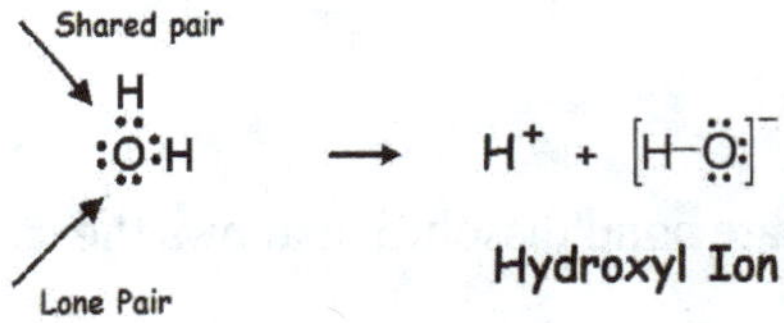

This is Coordinate Covalent Bond.

Applications

Understanding how coordinate bonds form can be instrumental in designing complex organic molecules.

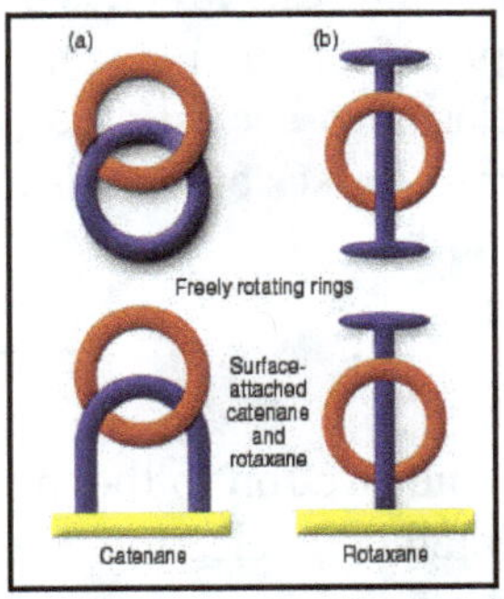

(a) Catenanes are formed from two or more molecular rings interlocked like links in a chain.
(b) rotaxanes are formed from a ring molecule threaded on an axle molecule capped by two molecular stoppers.

Rotaxanes, shown in the righthand column, are one of many mechanically interlocked molecular shapes.

Rotaxanes are an example of a mechanically interlocked molecule, a complex organic molecule that is physically joined. These molecules are interesting to synthetic chemists as "potential building blocks for future molecular-scale devices—motors, sensors, and machines on the nanometer scale".

The image below shows an organic molecule that can self-assemble to form a rotaxane because the molecule was synthesized with coordinate bonds in targeted places. Those bonds are then broken and re-formed to give the molecule its desired shape.

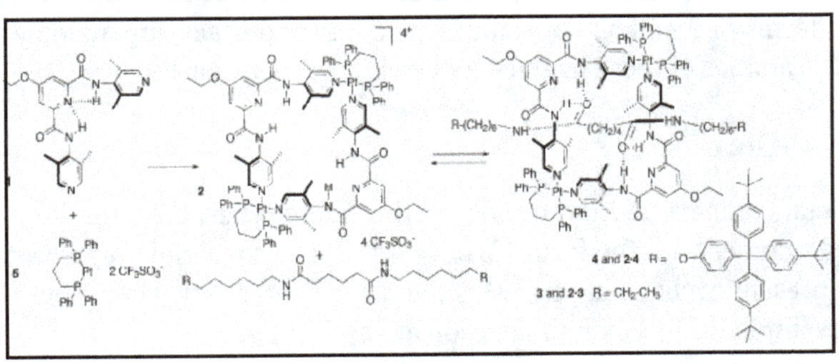

Precursor molecules to a rotaxane shape.

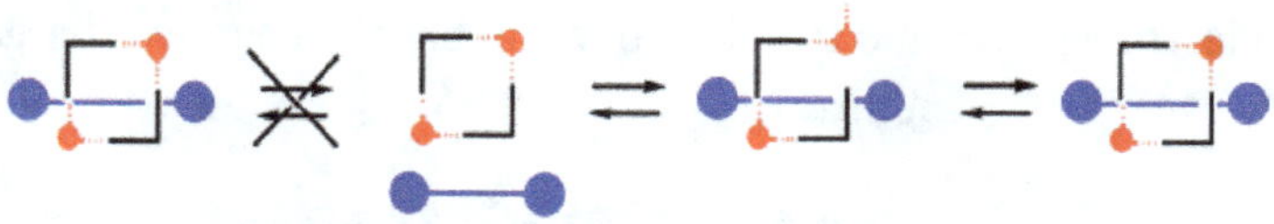

Schematic showing how coordinate bond dissolution allows the rotaxane to self-assemble.

Covalent Bond Classification (LXZ Notation)

The covalent bond classification (CBC) method is also referred to as the LXZ notation. It was published by M. L. H. Green in the mid-1990s as a solution for the need to describe covalent compounds such as organometallic complexes in a way that is not prone to limitations resulting from the definition of oxidation state. Instead of simply assigning a charge to an atom in the molecule (i.e. the oxidation state), the covalent bond classification method analyzes the nature of the ligands surrounding the atom of interest, which is often a transition metal. According to this method, there are three basic types of interactions that allow for coordination of the ligand. The three types of interaction are classified according to whether the ligating group donates two, one, or zero electrons. These three classes of ligands are respectively given the symbols L, X and Z.

Types of Ligands

X-type ligands are those that donate one electron to the metal and accept one electron from the metal when using the neutral ligand method of electron counting, or donate two electrons to the metal when using the donor pair method of electron counting. Regardless of whether it is considered neutral or anionic, these ligands yield normal covalent bonds. A few examples of this type of ligand are H, halogens (Cl, Br, F, etc.), OH, CN, CH_3, and NO.

L-type ligands are neutral ligands that donate two electrons to the metal center regardless of the electron counting method being used. These electrons can come from lone pairs, pi or sigma donors. The bonds formed between these ligands and the metal are dative covalent bonds, which are also known as coordinate bonds. Examples of this type of ligand include CO, PR_3, NH_3, H_2O, carbenes (=CRR'), and alkenes.

Z-type ligands are those that accept two electrons from the metal center as opposed to the donation occurring with the other two types of ligands. However, these ligands also form dative covalent bonds like the L-type. This type of ligand is not usually used, because in certain situations it can be written in terms of L and X. For example, if a Z ligand is accompanied by an L type, it can be written as X_2. Examples of these ligands are Lewis acids, such as BR_3.

Uses of the Notation

When given a metal complex and the trends for the ligand types, the complex can be written in a more simplified manner with the form $[ML_lX_xZ_z]^{Q\pm}$. The subscripts represent the numbers of each ligand type present in that complex, M is the metal center and Q is the overall charge on the complex. Some examples of this overall notation are as follows:

Condensed formula	LXZ notation
$[Mn(CO)_6]^+$	$[ML_6]^+$
$[Ir(CO)(PPh_3)_2(Cl)(NO)]^{2+}$	$[ML_3X_2]^{2+}$
$[Fe(CO)_2(CN)_4]^{2-}$	$[ML_2X_4]^{2-}$

Also from this general form, the values for electron count, oxidation state, coordination number, number of d-electrons, valence number and the ligand bond number can be calculated.

Electron Count = $N+x+2l-Q$

Where, N is the group number of the metal.

Oxidation State (OS) = $x+Q$

Coordination Number (CN) = $x+l$

Number of d-electrons (dn) = $N-OS$

= $N-(x+Q)$

Valence Number (VN) = $x+2z$ S

Ligand Bond Number (LBN) = $l+x+z$

Other Uses

This template of writing a metal complex also allows for better comparison of molecules with different charges. This can happen when the assignment is reduced to its "equivalent neutral class". The equivalent neutral class is the classification of the complex if the charge was localized on the ligand as opposed to the metal center. In other words, the equivalent neutral class is the representation of the complex as though there was no charge.

Atomic Radius

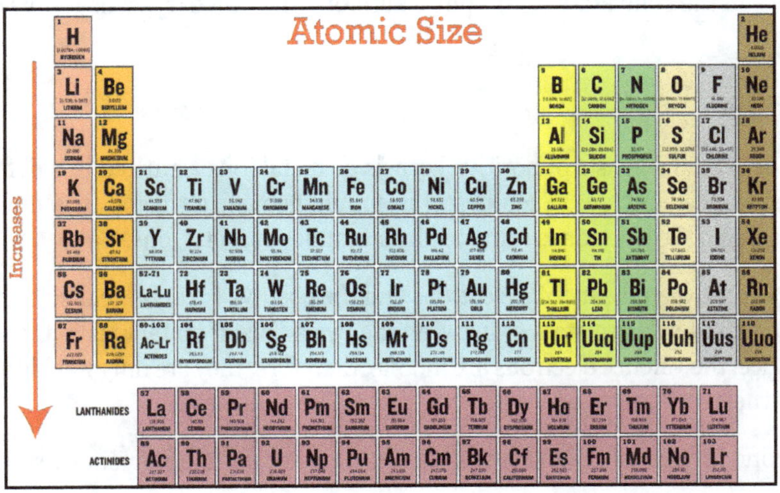

Atomic radius is the distance from the center of the nucleus to the outermost shell containing electrons. In other words, it is the distance from the center of the nucleus to the point up to which the density of the electron cloud is maximum.

Types of Atomic Radii

Atomic radii are divided into three types:

- Covalent radius
- Van der Waals radius
- Metallic radius

Therefore, we will study these three types of radius because they are vital for the better understanding of the subject.

Covalent Radius

Covalent radius is one half the distance between the nuclei of two covalently bonded atoms of the same element in a molecule. Therefore, $r_{covalent}$ = ½ (inter-nuclear distance between two bonded atoms). The inter-nuclear distance between two bonded atoms is called the bond length.

Therefore,

$$r_{covalent} = \frac{1}{2}(\text{bond length})$$

Van der Waals Radius

It is one half the distance between the nuclei of two identical non-bonded isolated atoms or two adjacent identical atoms belonging to two neighboring molecules of an element in the solid state. The magnitude of the Van der Waals radius is dependent on the packing of the atoms when the element is in the solid state.

For example, the inter-nuclear distance between two adjacent chlorine atoms of the two neighboring molecules in the solid state is 360 pm. Therefore, the Van der Waals radius of the chlorine atom is 180 pm.

Metallic Radius

A metal lattice or crystal consists of positive kernels or metal ions arranged in a definite pattern in a sea of mobile valence electrons. Each kernel is simultaneously attracted by a number of mobile electrons and each mobile electron is attracted by a number of metal ions.

Force of attraction between the mobile electrons and the positive kernels is called the metallic bond. It is one half the inter-nuclear distances between the two adjacent metal ions in the metallic lattice. In a metallic lattice, the valence electrons are mobile, therefore, they are only weakly attracted by the metal ions or kernels.

In a covalent bond, a pair of electrons is strongly attracted by the nuclei of two atoms. Thus, a metallic radius is always longer than its covalent radius. For example, the metallic radius of sodium is 186 pm whereas its covalent radius as determined by its vapor which exists as Na_2 is 154 pm. The metallic radius of Potassium is 231 pm while its covalent radius is 203 pm.

Variation of Atomic Radii in the Periodic Table

Variation within a Period

- The Covalent and Van der Waals radii decrease with increase in atomic number as we move from left to right in a period. The alkali metals at the extreme left of the periodic table have the largest size in a period. The halogens at the extreme right of the periodic table have the smallest size. The atomic size of nitrogen is the smallest. After nitrogen, atomic size increases for Oxygen and then decreases for fluorine. The size of atoms of inert gases is larger than those of the preceding halogens.
- As we move from left to right in a period, nuclear charge increases by 1 unit in each succeeding element while the number of shells remains the same. This enhanced nuclear charge pulls the electrons of all the shells closer to the nucleus. This makes each individual shells smaller and smaller. This result in a decrease of the atomic radius as we move from left to right in a period.
- The atomic radius abruptly increases as we move from halogens to the inert gas. This is because inert gases have completely filled orbitals. Hence, the inter-electronic are maximum. We express the atomic size in terms of Van der Waals radius since they do not form covalent bonds. Van der Waals radius is larger than the covalent radius. Therefore, the atomic size of an inert gas in a period is much higher than that of preceding halogen.

Variation Within a Group

The atomic radii of elements increase with an increase in atomic number from top to bottom in a group. As we move down the group, the principal quantum number increases. A new energy shell is added at each succeeding element. The valence electrons lie farther and farther away from the nucleus. As a result, the attraction of the nucleus for the electron decreases. Hence, the atomic radius increases.

Functional Group

Functional groups in organic chemistry are those which are responsible for the changes in the physical and chemical properties of an organic compound. There are various such groups in organic chemistry. It can be defined as:

> "An atom or group of atoms which show a characteristic unique reactivity when treated with some reagents and also which impart some unique chemical, as well as physical properties to the primary molecule, is defined as a functional group."

Molecules are divided into two types in organic chemistry:

- Organic molecules
- Inorganic molecules

The molecules which consist of carbon atoms along with hydrogen atoms are called as organic molecules like hydrocarbons, and the functional groups are mainly dealt with the organic molecules since they exhibit non-polarity kind of nature and also due to the high electro-negativity. These groups also decide the properties and the functionality of a chemical compound. Some of the examples of functional groups are carboxylic acid (RCOO-) and ester (R-COO-R). This comprises of an acyl group (RCO-) and an alkoxy group (-OR').

Types of Functional Groups

A variety of functional groups exist in organic chemistry. Most of them are carbon-based compounds. Some groups form a single bond with most of the hetero-atoms such as oxygen, iodine, nitrogen, sulfur, etc. There is also some other kind of groups having multiple bonds with the hetero-atoms such as oxygen, nitrogen atoms, etc.

The lists of some of the common functional groups are contained in the table below:

Group formula	Class name	Example	Common name
C=C	Alkene	CH_2=CH_2	Ethylene
Arene	C_6H_6	Benzene	
-OH	Alcohol	CH_3OH	Wood alcohol
-Cl	Chloro	CH_3Cl	Methyl chloride
C-X	Halide	CH_3-I	Methyl iodide
C-O-C	Ether	CH_3-CH_2-O-CH_2-CH_3	Ether
C-N	Amine	CH_3-NH_2	Methyl amine
Aldehyde	CH_3-CHO	Acetaldehyde	
Ketone	CH_3-CO-CH_3	Acetone	
Carboxylic acid	CH_3-COOH	Acetic acid	
Ester	CH_3-CO_2-CH_2-CH_3	Ethyl acetate	

The 'default' in organic chemistry (essentially, the *lack* of any functional groups) is given the term alkane, characterized by single bonds between carbon and carbon, or between carbon and hydrogen. Methane, CH_4, is the natural gas you may burn in your furnace. Octane, C_8H_{18}, is a component of gasoline.

Alkanes

Alkenes (sometimes called olefins) have carbon-carbon double bonds, and alkynes have carbon-carbon triple bonds. Ethene, the simplest alkene example, is a gas that serves as a cellular signal in fruits to stimulate ripening. (If you want bananas to ripen quickly, put them in a paper bag along with an apple - the apple emits ethene gas, setting off the ripening process in the bananas). Ethyne, commonly called acetylene, is used as a fuel in welding blow torches.

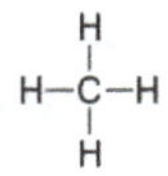

methane

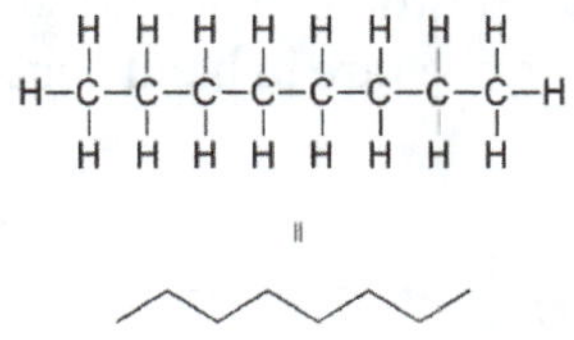

octane

Alkenes and Alkynes

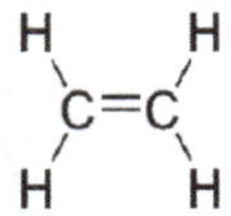

ethene
(an alkene)

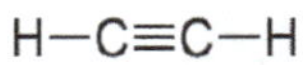

ethyne
(an alkyne)

Furthermore, many alkenes can take two geometric forms: cis or trans, The cis and trans forms of a given alkene are different molecules with different physical properties because, there is a very high energy barrier to rotation about a double bond. In the example below, the difference between cis and trans alkenes is readily apparent.

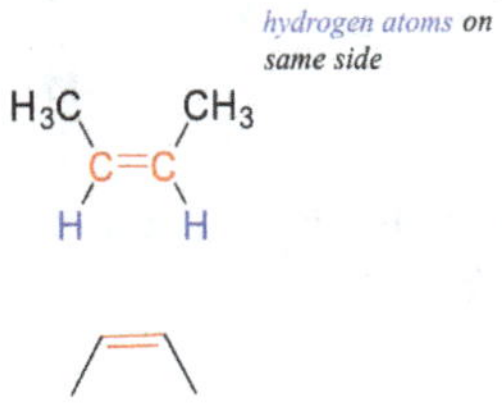

cis-alkene

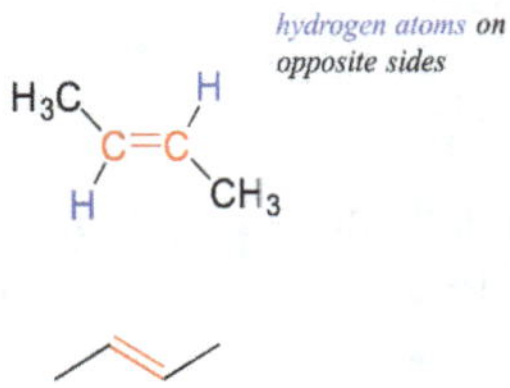

trans-alkene

Alkanes, alkenes, and alkynes are all classified as hydrocarbons, because they are composed solely of carbon and hydrogen atoms. Alkanes are said to be saturated hydrocarbons, because the carbons are bonded to the maximum possible number of hydrogen- in other words, they are saturated with hydrogen atoms. The double and triple-bonded carbons in alkenes and alkynes have fewer hydrogen atoms bonded to them, they are thus referred to as unsaturated hydrocarbons. Hydrogen can be added to double and triple bonds, in a type of reaction called 'hydrogenation'.

The aromatic group is exemplified by benzene (which used to be a commonly used solvent on the organic lab, but which was shown to be carcinogenic), and naphthalene, a compound with a distinctive ‹mothball' smell. Aromatic groups are planar (flat) ring structures, and are widespread in nature.

Aromatics

When the carbon of an alkane is bonded to one or more halogens, the group is referred to as a alkyl halide or haloalkane. Chloroform is a useful solvent in the laboratory, and was one of the earlier anesthetic drugs used in surgery. Chlorodifluoromethane was used as a refrigerant and in aerosol sprays until the late twentieth century, but its use was discontinued after it was found to have harmful effects on the ozone layer. Bromoethane is a simple alkyl halide often used in organic synthesis. Alkyl halides groups are quite rare in biomolecules.

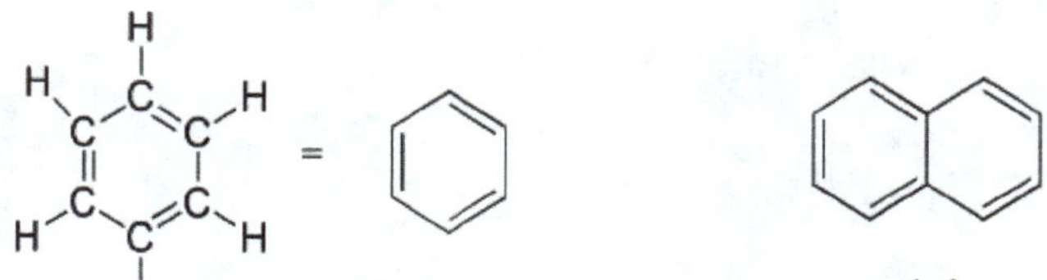

benzene napthalene

Haloalkanes

In the alcohol functional group, a carbon is single-bonded to an OH group (the OH group, by itself, is referred to as a hydroxyl). Except for methanol, all alcohols can be classified as primary, secondary, or tertiary. In a primary alcohol, the carbon bonded to the OH group is also bonded to only one other carbon. In a secondary alcohol and tertiary alcohol, the carbon is bonded to two or three other carbons, respectively. When the hydroxyl group is directly attached to an aromatic ring, the resulting group is called a phenol. The sulfur analog of an alcohol is called a thiol.

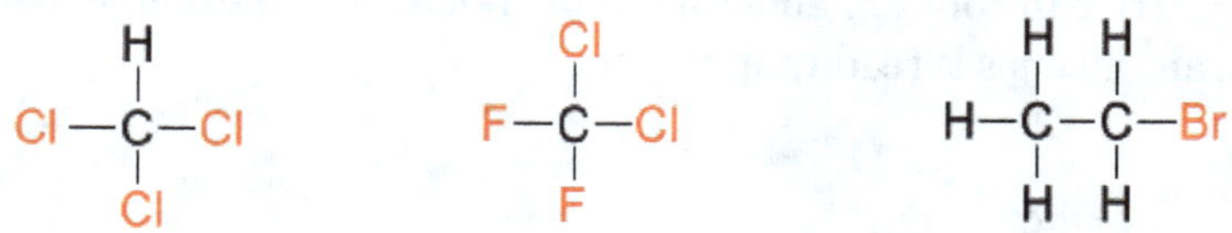

trichloromethane (chloroform) dichlorodifluoromethane (Freon-12) bromoethane

Alcohols, Phenols and Thiols

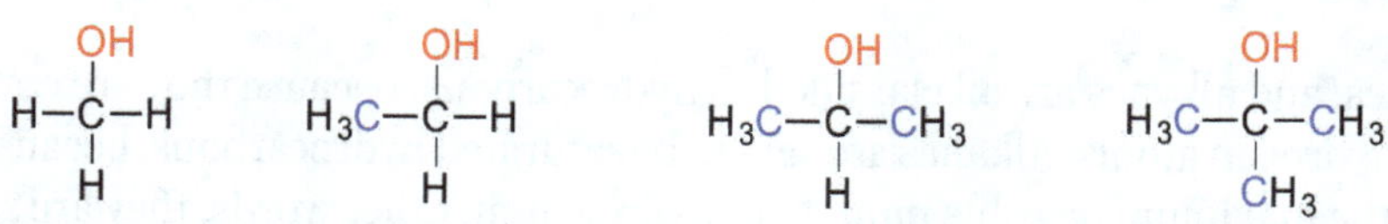

methanol primary alcohol secondary alcohol tertiary alcohol

phenol thiol

Note that the definition of a phenol states that the hydroxyl oxygen must be *directly* attached to one of the carbons of the aromatic ring. The compound below, therefore, is *not* a phenol - it is a primary alcohol.

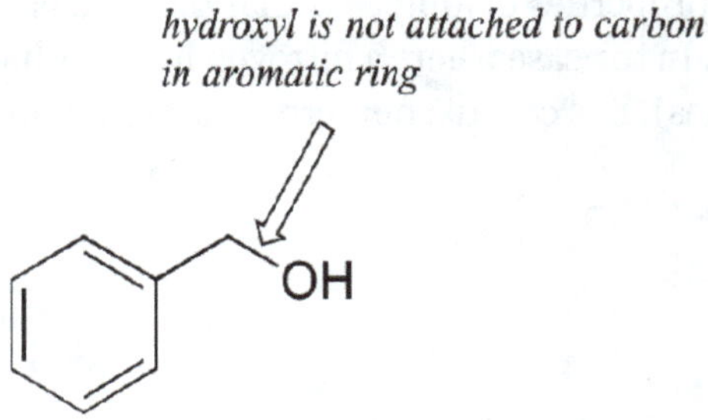

primary alcohol
(*not* a phenol)

The distinction is important, because as we will see later, there is a significant difference in the reactivity of alcohols and phenols.

The deprotonated forms of alcohols, phenols, and thiols are called alkoxides, phenolates, and thiolates, respectively. A protonated alcohol is an oxonium ion.

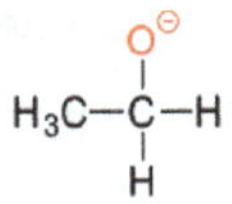

alkoxide

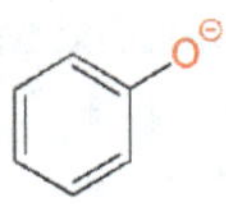

phenolate

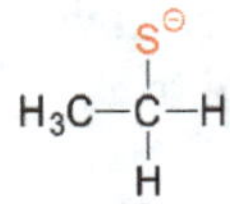

thiolate

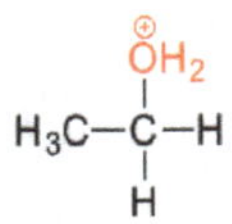

oxonium

In an ether functional group, central oxygen is bonded to two carbons. Below is the structure of diethyl ether, a common laboratory solvent and also one of the first compounds to be used as an anesthetic during operations. The sulfur analog of an ether is called a thioether or sulfide.

Ethers and Sulfides

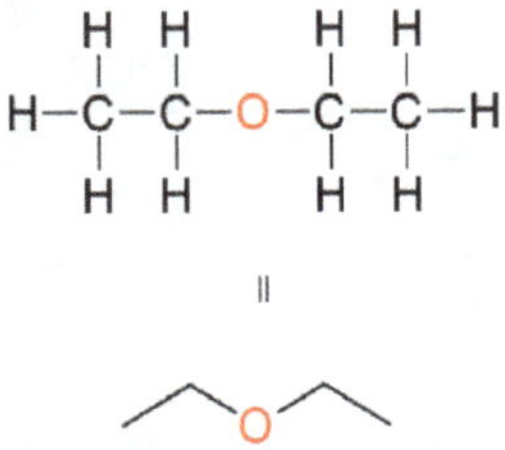

an ether

a sulfide

Amines are characterized by nitrogen atoms with single bonds to hydrogen and carbon. Just as there are primary, secondary, and tertiary alcohols, there are primary, secondary, and tertiary amines. Ammonia is a special case with no carbon atoms.

One of the most important properties of amines is that they are basic, and are readily protonated to form ammonium cations. In the case where a nitrogen has four bonds to carbon (which is somewhat unusual in biomolecules), it is called a quaternary ammonium ion.

Amines

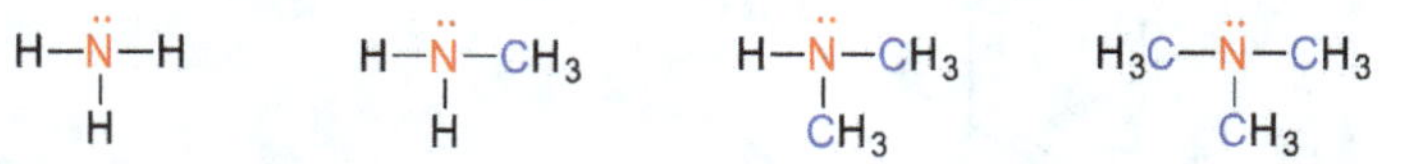

ammonia a primary amine a secondary amine a tertiary amine

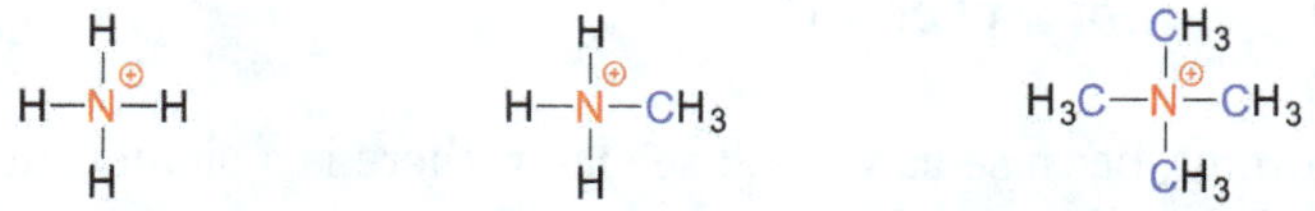

ammonium ion a primary ammonium ion a quaternary ammonium ion

There is a difference in the usage of the terms 'primary', 'secondary', and 'tertiary' when they are applied to alcohols and amines for the definitions are different. In alcohols, what matters is how many other carbons, the alcohol *carbon* is bonded to, while in amines, what matters is how many carbons the *nitrogen* is bonded to.

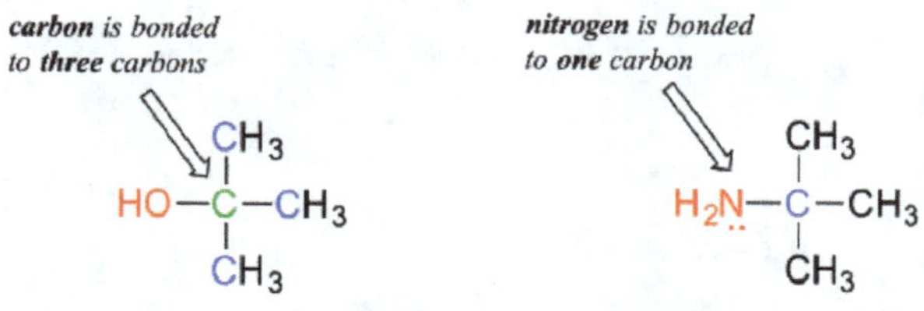

a *tertiary* alcohol a *primary* amine

Phosphate and its derivative functional groups are ubiquitous in biomolecules. Phosphate linked to a single organic group is called a phosphate ester; when it has two links to organic groups it is called a phosphate diester. A linkage between two phosphates creates a phosphate anhydride.

Organic phosphates

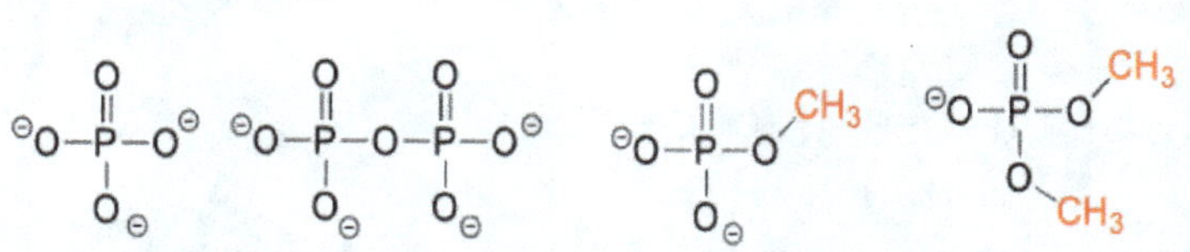

phosphate a phosphate anhydride a phosphate ester a phosphate diester

There are a number of functional groups that contain a carbon-oxygen double bond, which is commonly referred to as a carbonyl. Ketones and aldehydes are two closely related carbonyl-based functional groups that react in very similar ways. In a ketone, the carbon atom of a carbonyl is bonded to two other carbons. In an aldehyde, the carbonyl carbon is bonded on one side to a hydrogen, and on the other side to a carbon. The exception to this definition is formaldehyde, in which the carbonyl carbon has bonds to two hydrogens.

A group with a carbon-nitrogen double bond is called an imine, or sometimes a Schiff base (in this book we will use the term 'imine').

Aldehydes, Ketones, and Imines

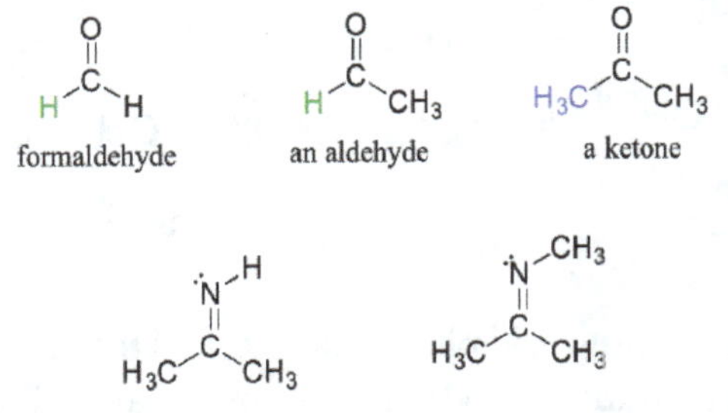

imines

When a carbonyl carbon is bonded on one side to a carbon (or hydrogen) and on the other side to an oxygen, nitrogen, or sulfur, the functional group is considered to be one of the 'carboxylic acid derivatives', a designation that describes a set of related functional groups. The eponymous member of this family is the carboxylic acid functional group, in which the carbonyl is bonded to a hydroxyl group. The conjugate base of a carboxylic acid is a carboxylate. Other derivatives are carboxylic esters (usually just called 'esters'), thioesters, amides, acyl phosphates, acid chlorides, and acid anhydrides. With the exception of acid chlorides and acid anhydrides, the carboxylic acid derivatives are very common in biological molecules and metabolic pathways.

Carboxylic Acid Derivatives

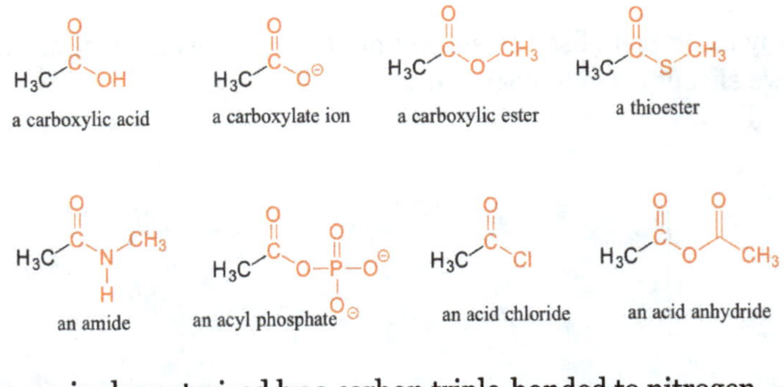

Finally, a nitrile group is characterized by a carbon triple-bonded to nitrogen.

Nitriles

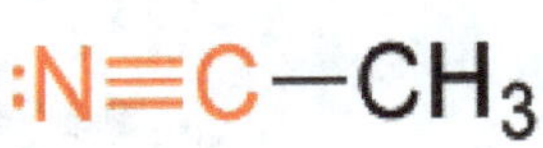

acetonitrile

A single compound often contains several functional groups, particularly in biological organic chemistry. The six-carbon sugar molecules glucose and fructose, for example, contain aldehyde and ketone groups, respectively, and both contain five alcohol groups (a compound with several alcohol groups is often referred to as a 'polyol').

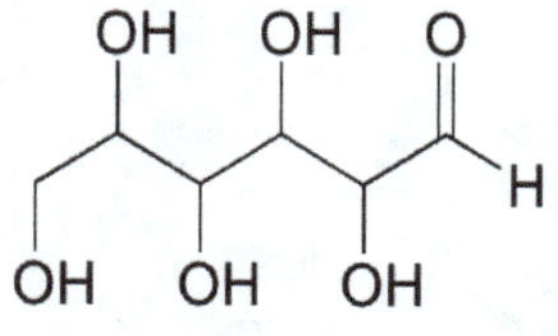

glucose

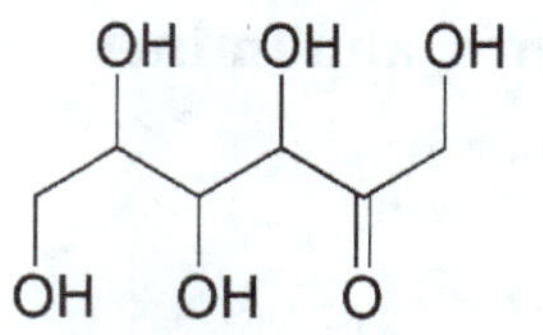

fructose

The hormone testosterone, the amino acid phenylalanine, and the glycolysis metabolite dihydroxyacetone phosphate all contain multiple functional groups, as labeled below:

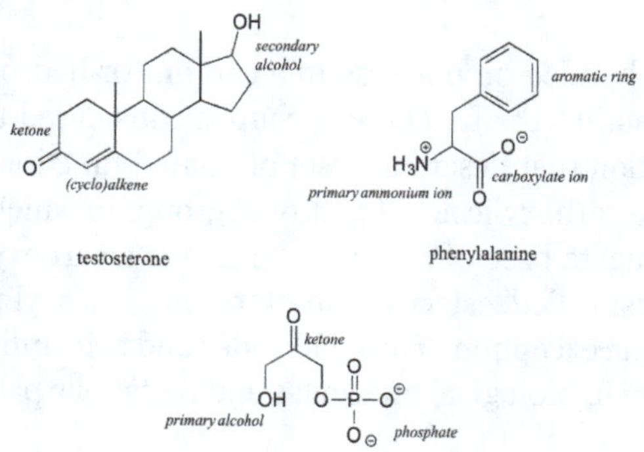

testosterone

phenylalanine

dihydroxyacetone phosphate (DHAP)

While not in any way a complete list, the subject matter here covers most of the important functional groups that we encounter in biological organic chemistry.

Hydrocarbons

Hydrocarbons are obtained by refining petroleum at refineries such as this one.

A hydrocarbon is any chemical compound that is constituted of just the elements carbon (C) and hydrogen (H). Each hydrocarbon molecule consists of a carbon backbone, or "carbon skeleton," with hydrogen atoms attached to that backbone.

Hydrocarbons are among the Earth's most important natural resources. They are currently the main source of the world's electric energy and heat energy (such as for heating buildings) because they produce large amounts of heat when burned. The gasoline that serves as fuel for automobiles consists primarily of hydrocarbons. In addition, many hydrocarbons serve as base materials for the synthesis of organic chemicals used in the production of consumer products and industrial materials.

Natural Occurrence and Extraction

Hydrocarbons are the main constituents of petroleum (literally, "rock oil"), also called "oil," and natural gas. They are commonly found in and extracted from the Earth's subsurface. Petroleum is a mixture of liquid hydrocarbons, while natural gas is mainly constituted of methane gas.

The extraction of liquid hydrocarbon fuel from a number of sedimentary basins has been integral to modern energy development. Hydrocarbons are mined from tar sands and oil shale. These reserves require distillation and upgrading to produce synthetic crude and petroleum. A future source of methane may be methane hydrates found on ocean floors.

Types of Hydrocarbons

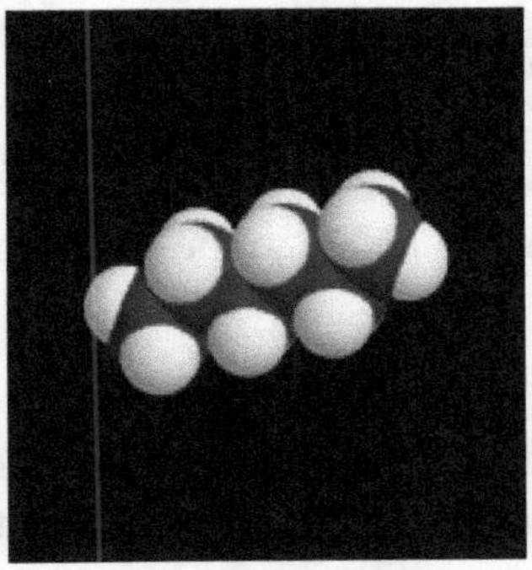	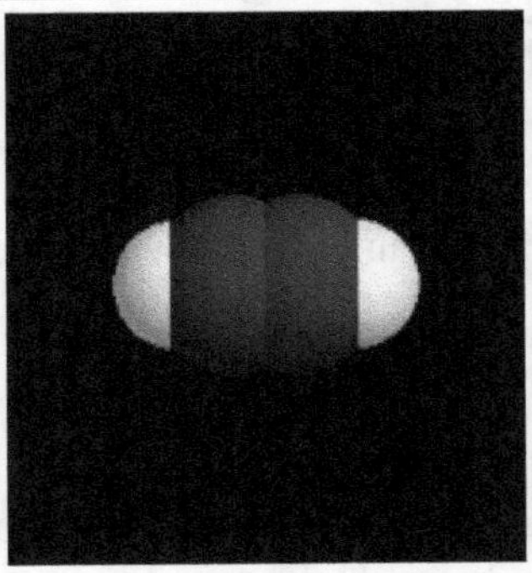	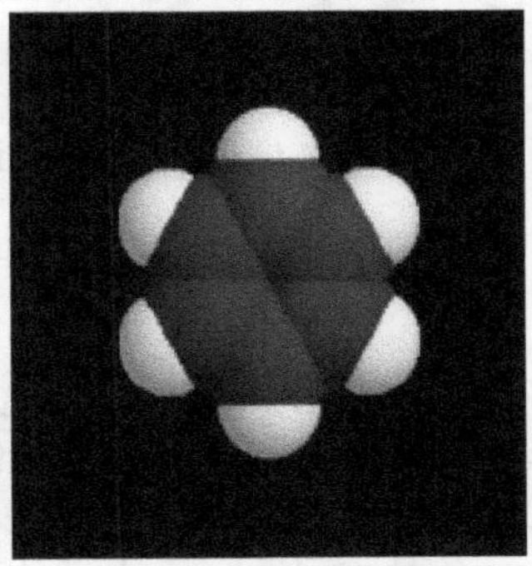
A model of the alkane known as *hexane.*	A model of the alkyne called *ethyne.*	A model of an arene called *benzene.*

There are essentially three types of hydrocarbons: Saturated hydrocarbons, also known as alkanes: In each molecule of an alkane, the chemical bonds that join the carbon atoms are single covalent bonds. If the alkane molecule includes a ring of carbon atoms (all connected by single covalent bonds), it is called a cycloalkane. Unsaturated hydrocarbons, which are subdivided into two groups:

- Alkenes: Each molecule of an alkene contains at least one double covalent bond between carbon atoms.
- Alkynes: Each molecule of an alkyne contains at least one triple covalent bond between carbon atoms.
- Aromatic hydrocarbons, or arenes: Each molecule of an aromatic hydrocarbon contains at least one aromatic ring, in which the bonds between carbon atoms are aromatic bonds.

When organic compounds are considered in general, saturated and unsaturated hydrocarbons are placed in the category known as aliphatic compounds, while aromatic hydrocarbons are categorized as aromatic compounds.

Some Simple Hydrocarbons

The simplest hydrocarbon is methane, the main constituent of natural gas. Its chemical formula, CH_4, indicates that each molecule of methane consists of four hydrogen atoms attached to one carbon atom. Ethane, with the formula C_2H_6, is a hydrocarbon (more specifically, an alkane) in which each molecule has two carbon atoms held together with a single covalent bond, and three hydrogen atoms are bound to each carbon atom. Each molecule of propane (C_3H_8) has three carbon atoms, and each molecule of butane (C_4H_{10}) has four carbons.

General Chemical Formulas

If a hydrocarbon molecule contains "n" carbon atoms, and the type of hydrocarbon is known, its general chemical formula can be written as follows:

- Alkane: C_nH_{2n+2}.
- Cycloalkane: C_nH_{2n} (assuming a single ring, with all the carbon atoms in the ring structure).
- Alkene: C_nH_{2n} (assuming only one double bond in each molecule).
- Alkyne: C_nH_{2n-2} (assuming only one triple bond in each molecule).
- Aromatic ring: C_nH_n (assuming a single ring, with all the carbon atoms in the ring structure).

The above chemical formulas are based on the assumption that each carbon atom forms four covalent bonds, including bonds with hydrogen atoms and other carbon atoms. A double covalent bond counts as two bonds; a triple covalent bond counts as three bonds. A carbon atom in an aromatic ring is a special case and can have only one hydrogen atom attached to it.

Molecular Graph

A "molecular graph" of a hydrocarbon is a simple representation of the carbon skeleton of the molecule under consideration. Each line in the molecular graph represents a covalent bond that connects two carbon atoms. Thus, although the graph shows no symbols for carbon atoms, carbon atoms are (by convention) located at each end of each line. The symbols for hydrogen atoms are not shown either, but each carbon atom holds as many hydrogen atoms as it needs to form four covalent bonds.

Uses of Hydrocarbons

Most hydrocarbons are combustible. When burned, they produce carbon dioxide, water, and large quantities of heat. This energy is used to heat homes and other buildings and to generate electricity. When heating a home, for example, oil or natural gas is burned and the energy released is used to heat water or air. The hot water or air is then circulated around the building.

A similar principle is used to create electric energy in power plants. A hydrocarbon fuel (such as natural gas) is burned, the energy released is used to convert water into steam, and the steam is used to drive turbines that generate electricity.

Hydrocarbons are also raw materials that serve as feedstock for the production of a wide range of organic chemicals, which in turn are used for such products as plastics, pigments, solvents, pharmaceuticals, and explosives.

Effects on Health and the Environment

Hydrocarbon vapors can be harmful if inhaled. Moreover, hydrocarbons contribute to the formation of ozone in the troposphere. For these reasons, hydrocarbons in the atmosphere are considered to be pollutants.

Ideally, the combustion of hydrocarbons should produce only carbon dioxide, water, and heat. Yet, incomplete combustion leads to the production of carbon monoxide, a toxic gas. Carbon monoxide binds to hemoglobin (in the blood) more readily than oxygen does; so, when carbon monoxide is inhaled, it blocks oxygen from being absorbed and leads to suffocation. Moreover, the hydrocarbon fuel being burned may contain other substances that are harmful when released.

Moreover, carbon dioxide, a product of the combustion of hydrocarbons, is a greenhouse gas—a gas that helps trap heat in the Earth's atmosphere. This has led to concerns that the excessive burning of hydrocarbon fuels contributes to global warming.

IUPAC Nomenclature of Organic Chemistry

The base part of the name reflects the number of carbons in what you have assigned to be the parent chain. The suffix of the name reflects the type(s) of functional group(s) present on (or within) the parent chain. Other groups which are attached to the parent chain are called substituents.

Alkanes - Saturated Hydrocarbons

The names of the straight chain saturated hydrocarbons for up to a 12 carbon chain are shown below. The names of the substituents formed by the removal of one hydrogen from the end of the chain is obtained by changing the suffix -ane to -yl.

Number of Carbons	Name
1	methane
2	ethane
3	propane
4	butane
5	pentane
6	hexane
7	heptane
8	octane

9	nonane
10	decane
11	undecane
12	dodecane

There are a few common branched substituents which you should memorize. These are shown below:

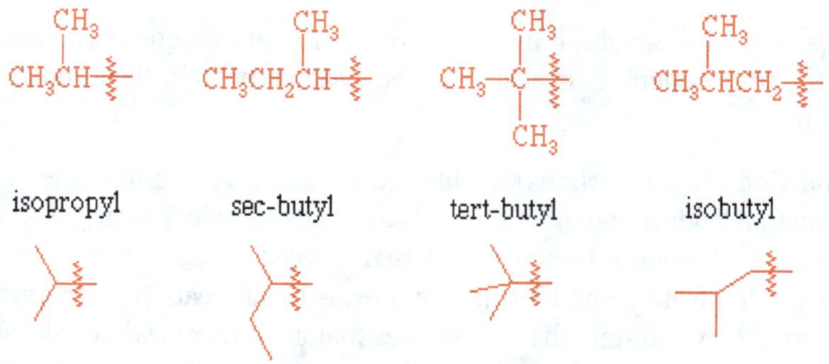

Here is a simple list of rules to follow, some examples are given at the end of the list:

- Identify the longest carbon chain. This chain is called the parent chain.
- Identify all of the substituents (groups appending from the parent chain).
- Number the carbons of the parent chain from the end that gives the substituents the lowest numbers. When compairing a series of numbers, the series that is the "lowest" is the one which contains the lowest number at the occasion of the first difference. If two or more side chains are in equivalent positions, assign the lowest number to the one which will come first in the name.
- If the same substituent occurs more than once, the location of each point on which the substituent occurs is given. In addition, the number of times the substituent group occurs is indicated by a prefix (di, tri, tetra, etc.).
- If there are two or more different substituents they are listed in alphabetical order using the base name (ignore the prefixes). The only prefix which is used when putting the substituents in alphabetical order is iso as in isopropyl or isobutyl. The prefixes sec- and tert- are not used in determining alphabetical order except when compared with each other.
- If chains of equal length are competing for selection as the parent chain, then the choice goes in series to:
    - The chain which has the greatest number of side chains;
    - The chain whose substituents have the lowest- numbers;
    - The chain having the greatest number of carbon atoms in the smaller side chain;
    - The chain having the least branched side chains.

- A cyclic (ring) hydrocarbon is designated by the prefix cyclo-which appears directly in front of the base name.

In summary, the name of the compound is written out with the substituents in alphabetical order followed by the base name (derived from the number of carbons in the parent chain). Commas are used between numbers and dashes are used between letters and numbers. There are no spaces in the name.

Here are some examples:

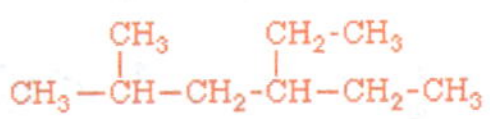

4-ethyl-2-methylhexane

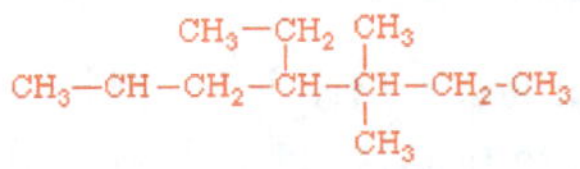

4-ethyl-3,3-dimethylheptane

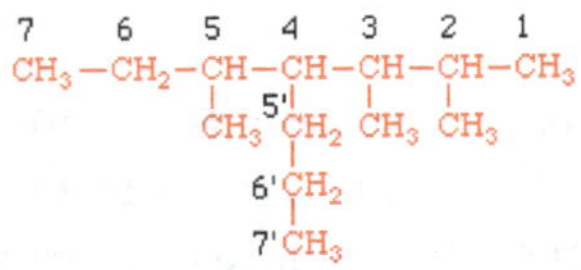

2,3,5-trimethyl-4-propylheptane
(NOT: 2,3-dimethyl-4-sec-butylheptane)

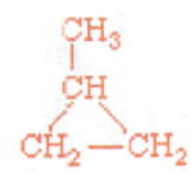

methylcyclopropane

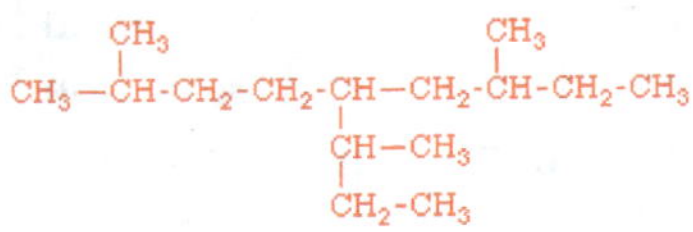

5-sec-butyl-2,7-dimethylnonane

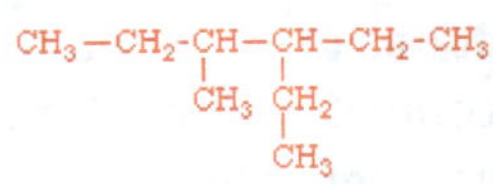

3-ethyl-4-methylhexane

Alkyl Halides

The halogen is treated as a substituent on an alkane chain. The halo- substituent is considered of equal rank with an alkyl substituent in the numbering of the parent chain. The halogens are represented as follows:

F	fluoro-
Cl	chloro-
Br	bromo-
I	iodo-

Here are some examples:

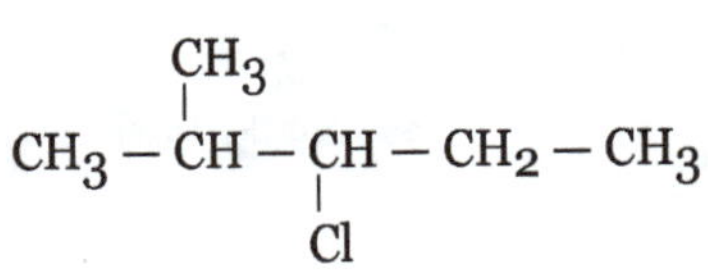

3-chloro-2-methylpentane

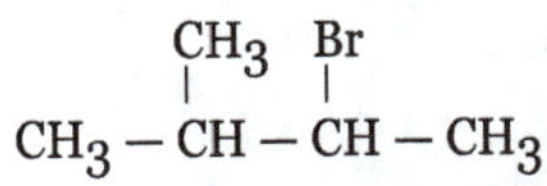

2-bromo-3-methylbutane

Alkenes and Alkynes - Unsaturated Hydrocarbons

Double bonds in hydrocarbons are indicated by replacing the suffix -anewith -ene. If there is more than one double bond, the suffix is expanded to include a prefix that indicates the number of double bonds present (-adiene, -atriene, etc.). Triple bonds are named in a similar way using the suffix -yne. The position of the multiple bond(s) within the parent chain is indicated by placing the number(s) of the first carbon of the multiple bond(s) directly in front of the base name.

Here is an important list of rules to follow:

- The parent chain is numbered so that the multiple bonds have the lowest numbers (double and triple bonds have priority over alkyl and halo substituents).
- When both double and triple bonds are present, numbers as low as possible are given to double and triple bonds even though this may at times give "-yne" a lower number than "-ene". When there is a choice in numbering, the double bonds are given the lowest numbers.
- When both double and triple bonds are present, the -en suffix follows the parent chain directly and the -yne suffix follows the -en suffix (notice that the e is left off, -en instead of -ene). The location of the double bond(s) is indicated before the parent name as before, and the location of the triple bond(s) is indicated between the -en and -yne suffixes.
- For a branched unsaturated acyclic hydrocarbon, the parent chain is the longest carbon chain that contains the maximum number of double and triple bonds. If there are two or more chains competing for selection as the parent chain (chain with the most multiple bonds), the choice goes to:
    - The chain with the greatest number of carbon atoms,
    - The carbon atoms being equal, the chain containing the maximum number of double bonds.
- If there is a choice in numbering not previously covered, the parent chain is numbered to give the substituents the lowest number at the first point of difference.

Here are some examples:

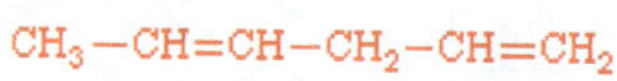

1,4-hexadiene

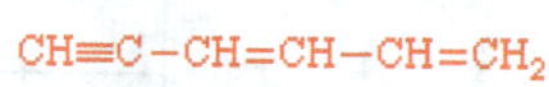

1,3-hexadien-5-yne

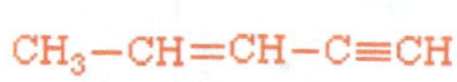

3-penten-1-yne

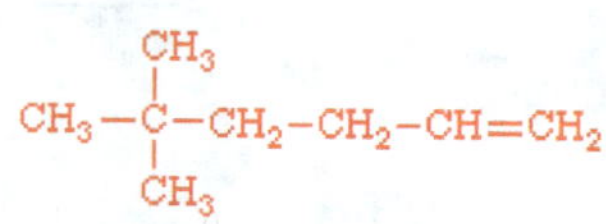

5,5-dimethyl-1-hexene

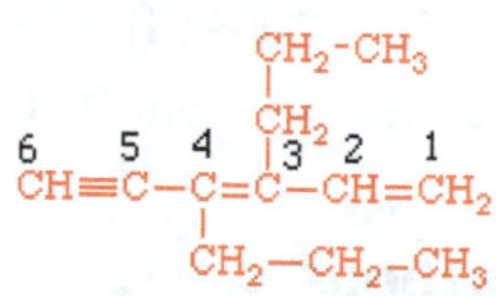

3,4-dipropyl-1,3-hexadien-5-yne

1,4,4-trimethylcyclobutene
(NOT: 2,3,3-trimethylcyclobutene)

Alcohols

Alcohols are named by replacing the suffix -ane with -anol. If there is more than one hydroxyl group (-OH), the suffix is expanded to include a prefix that indicates the number of hydroxyl groups present (-anediol, -anetriol, etc.). The position of the hydroxyl group(s) on the parent chain is indicated by placing the number(s) corresponding to the location(s) on the parent chain directly in front of the base name (same as alkenes).

Here is an important list of rules to follow:

- The hydroxyl group takes precedence over alkyl groups and halogen substituents, as well as double bonds, in the numbering of the parent chain.
- When both double bonds and hydroxyl groups are present, the -en suffix follows the parent chain directly and the -ol suffix follows the -en suffix (notice that the e is left off, -en instead of -ene). The location of the double bond(s) is indicated before the parent name as before, and the location of the hydroxyl group(s) is indicated between the -en and -ol suffixes. See below for examples. Again, the hydroxyl gets priority in the numbering of the parent chain.
- If there is a choice in numbering not previously covered, the parent chain is numbered to give the substituents the lowest number at the first point of difference.

Here are some examples:

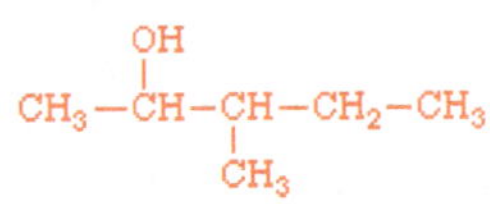

3-methyl-2-pentanol

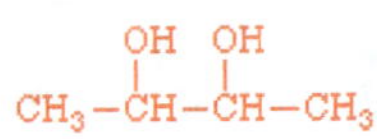

2,3-butanediol

2-cyclopenten-1-ol

Ethers

You are only expected to know how to name ethers by their commmon names. The two alkyl groups attached to the oxygen are put in alphabetical order with spaces between the names and they are followed by the word ether. The prefix di- is used if both alkyl groups are the same.

Here are some examples:

$$CH_3 - CH_2 - O - CH_2 - CH_3$$

diethyl ether

$$CH_3 - O - CH_2 - CH_3$$

ethyl methyl ether

Aldehydes

Aldehydes are named by replacing the suffix -ane with -anal. If there is more than one -CHO group, the suffix is expanded to include a prefix that indicates the number of -CHO groups present

(-anedial - there should not be more than 2 of these groups on the parent chain as they must occur at the ends). It is not necessary to indicate the position of the -CHO group because this group will be at the end of the parent chain and its carbon is automatically assigned as C-1.

Here is an important list of rules to follow:

- The carbonyl group takes precedence over alkyl groups and halogen substituents, as well as double bonds, in the numbering of the parent chain.
- When both double bonds and carbonyl groups are present, the -en suffix follows the parent chain directly and the -al suffix follows the -en suffix (notice that the e is left off, -en instead of -ene). The location of the double bond(s) is(are) indicated before the parent name as before, and the -al suffix follows the -en suffix directly. Remember it is not necessary to specify the location of the carbonyl group because it will automatically be carbon 1. Again, the carbonyl gets priority in the numbering of the parent chain.
- There are a couple of common names which are acceptable as IUPAC names. They are shown in the examples at the end of this list but at this point these names will not be accepted by the computer. Eventually they will be accepted.
- If there is a choice in numbering not previously covered, the parent chain is numbered to give the substituents the lowest number at the first point of difference.

Here are some examples:

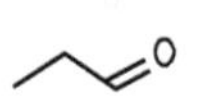

propanal

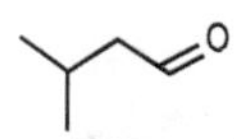

3-methylbutanal

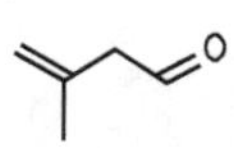

3-methyl-3-butanal

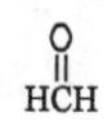

methanal
(common name: formaldehyde)

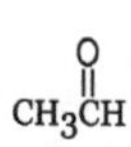

ethanal
(common name: acetaldehyde)

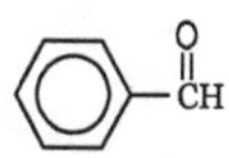

Benzaldehyde

Ketones

Ketones are named by replacing the suffix -ane with -anone. If there is more than one carbonyl group (C=O), the suffix is expanded to include a prefix that indicates the number of carbonyl groups present (-anedione, -anetrione, etc). The position of the carbonyl group(s) on the parent chain is indicated by placing the number(s) corresponding to the location(s) on the parent chain directly in front of the base name (same as alkenes).

Here is an important list of rules to follow:

- The carbonyl group takes precedence over alkyl groups and halogen substituents, as well as double bonds, in the numbering of the parent chain.

- When both double bonds and carbonyl groups are present, the -en suffix follows the parent chain directly and the -one suffix follows the -en suffix (notice that the e is left off, -en instead of -ene). The location of the double bond(s) is indicated before the parent name as before, and the location of the carbonyl group(s) is(are) indicated between the -en and -one suffixes.
- If there is a choice in numbering not previously covered, the parent chain is numbered to give the substituents the lowest number at the first point of difference.

Here are some examples,

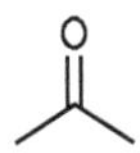

propanone
(common name : acetone

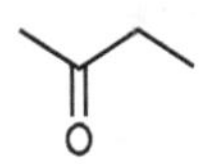

2-butanone

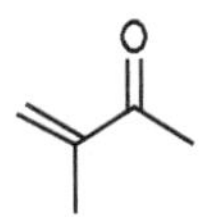

3-methyl-3-buten-2-one

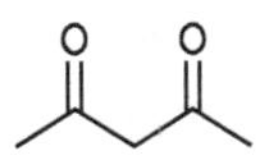

2, 4-pentanedione

Carboxylic Acids

Carboxylic acids are named by counting the number of carbons in the longest continuous chain including the carboxyl group and by replacing the suffix -ane of the corresponding alkane with -anoic acid. If there are two -COOH groups, the suffix is expanded to include a prefix that indicates the number of -COOH groups present (-anedioic acid - there should not be more than 2 of these groups on the parent chain as they must occur at the ends). It is not necessary to indicate the position of the -COOH group because this group will be at the end of the parent chain and its carbon is automatically assigned as C-1.

Here is an important list of rules to follow:

- The carboxyl group takes precedence over alkyl groups and halogen substituents, as well as double bonds, in the numbering of the parent chain.
- If the carboxyl group is attached to a ring the parent ring is named and the suffix -carboxylic acid is added.
- When both double bonds and carboxyl groups are present, the -en suffix follows the parent chain directly and the -oic acid suffix follows the -en suffix (notice that the e is left off, -en instead of -ene). The location of the double bond(s) is indicated before the parent name as before, and the -oic acid suffix follows the -en suffix directly. Remember it is not necessary to specify the location of the carboxyl group because it will automatically be carbon.
- There are several common names which are acceptable as IUPAC names. They are shown in the examples at the end of this list butat this point these names will not be accepted by the computer. Eventually they will be accepted.

- If there is a choice in numbering not previously covered, the parent chain is numbered to give the substituents the lowest number at the first point of difference.

Here are some examples:

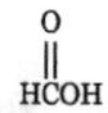

methanoic acid
(common name : formic acid)

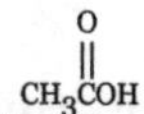

ethanoic acid
(common name : acetic acid)

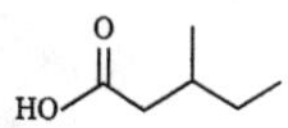

3-methylpentanoic acid

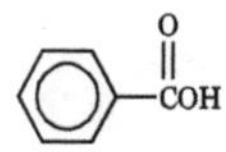

benzoic acid

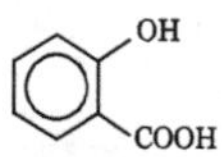

salicylic acid
(common name)

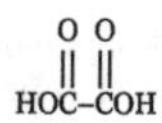

ethanedioic acid
(common name : oxalic acid)

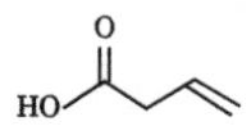

3-butenoic acid

Esters

Systematic names of esters are based on the name of the corresponding carboxylic acid. Remember esters look like this:

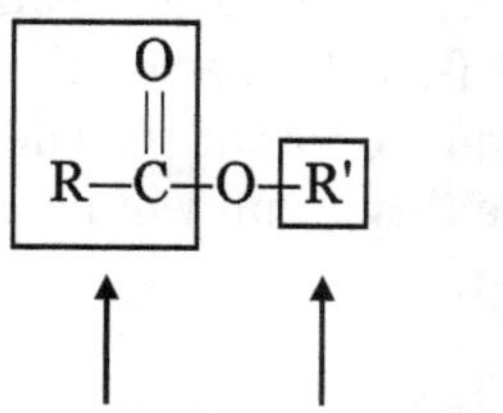

The alkyl group is named like a substituent using the -yl ending. This is followed by a space. The acyl portion of the name (what is left over) is named by replacing the -ic acid suffix of the corresponding carboxylic acid with -ate.

Here are some examples:

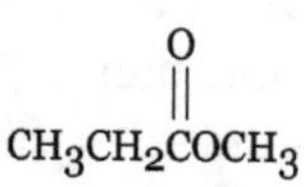

methyl propanoate

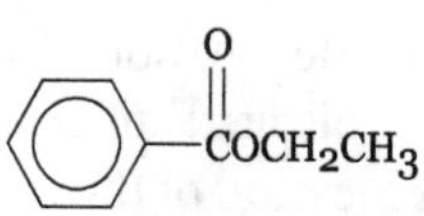

ethyl benzoate

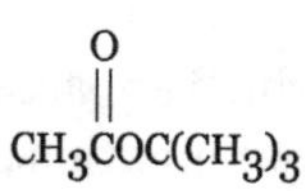

tert-butyl acetate

Amines

You are only expected to know how to name amines by their common names. They are named like ethers, the alkyl (R) groups attached to the nitrogen are put in alphabetical order with no spaces between the names and these are followed by the word amine. The prefixes di- and tri- are used if two or three of the alkyl groups are the same.

Note: Some books put spaces between the parts of the name, but we will not.

Here are some examples:

$NH(CH_3)_2$

dimethylamine

CH_3CH_2NH
|
CH_3

ethylmethylamine

References

- Formation-of-sigma-a-and-pi-n-bonds: chemistry-assignment.com, Retrieved 15, August 2020
- March, Jerry (1992). Advanced Organic Chemistry: Reactions, Mechanisms, and Structure. John Wiley & Sons. ISBN 0-471-60180-2
- Hydrocarbon: newworldencyclopedia.org, Retrieved 13, June 2020
- Langmuir, Irving (1919-06-01). "The Arrangement of Electrons in Atoms and Molecules". Journal of the American Chemical Society. 41 (6): 868–934. doi:10.1021/ja02227a002
- Properties-of-aromatic-compounds: lumenlearning.com, Retrieved 22, April 2020
- Campbell, Neil A.; Williamson, Brad; Heyden, Robin J. (2006). Biology: Exploring Life. Boston, MA: Pearson Prentice Hall. ISBN 0-13-250882-6.
- Covalent-bond, science: britannica.com, Retrieved 28, June 2020

6

Acid-base Chemistry: Concepts and Theories

An acid is hydrogen-containing molecule which has the ability to donate a proton to another molecule while base is any molecule which is capable of accepting a hydrogen ion from acid. Acid-base reaction is a chemical reaction in which exchange of hydrogen ions takes place between an acid and a base. This chapter delves into the diverse concepts and theories of acid-base chemistry.

Acid

An acid is a molecule or ion capable of donating a hydron (proton or hydrogen ion H^+), or, alternatively, capable of forming a covalent bond with an electron pair (a Lewis acid).

Zinc, a typical metal, reacting with hydrochloric acid, a typical acid.

The first category of acids is the proton donors or Brønsted acids. In the special case of aqueous solutions, proton donors form the hydronium ion H_3O^+ and are known as Arrhenius acids. Brønsted and Lowry generalized the Arrhenius theory to include non-aqueous solvents. A Brønsted or Arrhenius acid usually contains a hydrogen atom bonded to a chemical structure that is still energetically favorable after loss of H^+.

Aqueous Arrhenius acids have characteristic properties which provide a practical description of an acid. Acids form aqueous solutions with a sour taste, can turn blue litmus red, and react with bases and certain metals (like calcium) to form salts. The word *acid* is derived from the Latin *acidus/acēre* meaning *sour*. An aqueous solution of an acid has a pH less than 7 and is colloquially also referred to as 'acid' (as in 'dissolved in acid'), while the strict definition refers only to the solute. A lower pH means a higher acidity, and thus a higher concentration of positive hydrogen ions in the solution. Chemicals or substances having the property of an acid are said to be acidic.

Common aqueous acids include hydrochloric acid (a solution of hydrogen chloride which is found in gastric acid in the stomach and activates digestive enzymes), acetic acid (vinegar is a dilute aqueous solution of this liquid), sulfuric acid (used in car batteries), and citric acid (found in citrus fruits). As these examples show, acids (in the colloquial sense) can be solutions or pure substances, and can be derived from acids (in the strict sense) that are solids, liquids, or gases. Strong acids and some concentrated weak acids are corrosive, but there are exceptions such as carboranes and boric acid.

The second category of acids are Lewis acids, which form a covalent bond with an electron pair. An example is boron trifluoride (BF_3), whose boron atom has a vacant orbital which can form a covalent bond by sharing a lone pair of electrons on an atom in a base, for example the nitrogen atom in ammonia (NH_3). Lewis considered this as a generalization of the Brønsted definition, so that an acid is a chemical species that accepts electron pairs either directly *or* by releasing protons (H^+) into the solution, which then accept electron pairs. However, hydrogen chloride, acetic acid, and most other Brønsted-Lowry acids cannot form a covalent bond with an electron pair and are therefore not Lewis acids. Conversely, many Lewis acids are not Arrhenius or Brønsted-Lowry acids. In modern terminology, an *acid* is implicitly a Brønsted acid and not a Lewis acid, since chemists almost always refer to a Lewis acid explicitly as *a Lewis acid*.

Definitions and Concepts

Modern definitions are concerned with the fundamental chemical reactions common to all acids.

Most acids encountered in everyday life are aqueous solutions, or can be dissolved in water, so the Arrhenius and Brønsted-Lowry definitions are the most relevant.

The Brønsted-Lowry definition is the most widely used definition; unless otherwise specified, acid-base reactions are assumed to involve the transfer of a proton (H^+) from an acid to a base.

Hydronium ions are acids according to all three definitions. Interestingly, although alcohols and amines can be Brønsted-Lowry acids, they can also function as Lewis bases due to the lone pairs of electrons on their oxygen and nitrogen atoms.

Arrhenius Acids

The Swedish chemist Svante Arrhenius attributed the properties of acidity to hydrogen ions (H^+) or protons in 1884. An Arrhenius acid is a substance that, when added to water, increases the concentration of H^+ ions in the water. Note that chemists often write $H^+(aq)$ and refer to the hydrogen ion when describing acid-base reactions but the free hydrogen nucleus, a proton, does not exist alone in water, it exists as the hydronium ion, H_3O^+. Thus, an Arrhenius acid can also be described as a substance that increases the concentration of hydronium ions when added to water. Examples include molecular substances such as HCl and acetic acid.

Svante Arrhenius.

An Arrhenius base, on the other hand, is a substance which increases the concentration of hydroxide (OH^-) ions when dissolved in water. This decreases the concentration of hydronium because the ions react to form H_2O molecules:

$$H_3O^+_{(aq)} + OH^-_{(aq)} \rightleftharpoons H_2O_{(l)} + H_2O_{(l)}$$

Due to this equilibrium, any increase in the concentration of hydronium is accompanied by a decrease in the concentration of hydroxide. Thus, an Arrhenius acid could also be said to be one that decreases hydroxide concentration, while an Arrhenius base increases it.

In an acidic solution, the concentration of hydronium ions is greater than 10^{-7} moles per liter. Since pH is defined as the negative logarithm of the concentration of hydronium ions, acidic solutions thus have a pH of less than 7.

Brønsted–Lowry Acids

Acetic acid, a weak acid, donates a proton (hydrogen ion, highlighted in green) to water in an equilibrium reaction to give the acetate ion and the hydronium ion. Red: oxygen, black: carbon, white: hydrogen.

While the Arrhenius concept is useful for describing many reactions, it is also quite limited in its scope. In 1923 chemists Johannes Nicolaus Brønsted and Thomas Martin Lowry independently recognized that acid-base reactions involve the transfer of a proton. A Brønsted-Lowry acid (or simply Brønsted acid) is a species that donates a proton to a Brønsted-Lowry base. Brønsted-Lowry acid-base theory has several advantages over Arrhenius theory. Consider the following reactions of acetic acid (CH_3COOH), the organic acid that gives vinegar its characteristic taste:

$$CH_3COOH + H_2O \rightleftharpoons CH_3COO^- + H_3O^+$$
$$CH_3COOH + NH_3 \rightleftharpoons CH_3COO^- + NH_4^+$$

Both teories easily describe the first reaction: CH_3COOH acts as an Arrhenius acid because it acts as a source of H_3O^+ when dissolved in water, and it acts as a Brønsted acid by donating a proton to water. In the second example CH_3COOH undergoes the same transformation, in this case donating a proton to ammonia (NH_3), but does not relate to the Arrhenius definition of an acid because the reaction does not produce hydronium. Nevertheless, CH_3COOH is both an Arrhenius and a Brønsted-Lowry acid.

Brønsted-Lowry theory can be used to describe reactions of molecular compounds in nonaqueous solution or the gas phase. Hydrogen chloride (HCl) and ammonia combine under several different conditions to form ammonium chloride, NH_4Cl. In aqueous solution HCl behaves as hydrochloric acid and exists as hydronium and chloride ions. The following reactions illustrate the limitations of Arrhenius's definition:

$$1.\ H_3O^+_{(aq)} + Cl^+_{(aq)} + NH_3 \rightarrow Cl^-_{(aq)} + NH^+_{4(aq)} + H_2O$$
$$2.\ HCl_{(benzene)} + NH_{3(benzene)} \rightarrow NH_4Cl_{(s)}$$
$$3.\ HCl_{(g)} + NH_{3(g)} \rightarrow NH_4Cl_{(s)}$$

As with the acetic acid reactions, both definitions work for the first example, where water is the solvent and hydronium ion is formed by the HCl solute. The next two reactions do not involve the formation of ions but are still proton-transfer reactions. In the second reaction hydrogen chloride and ammonia (dissolved in benzene) react to form solid ammonium chloride in a benzene solvent and in the third gaseous HCl and NH_3 combine to form the solid.

Lewis Acids

A third, only marginally related concept was proposed in 1923 by Gilbert N. Lewis, which includes reactions with acid-base characteristics that do not involve a proton transfer. A Lewis acid is a species that accepts a pair of electrons from another species; in other words, it is an electron pair acceptor. Brønsted acid-base reactions are proton transfer reactions while Lewis acid-base reactions are electron pair transfers. Many Lewis acids are not Brønsted-Lowry acids. Contrast how the following reactions are described in terms of acid-base chemistry:

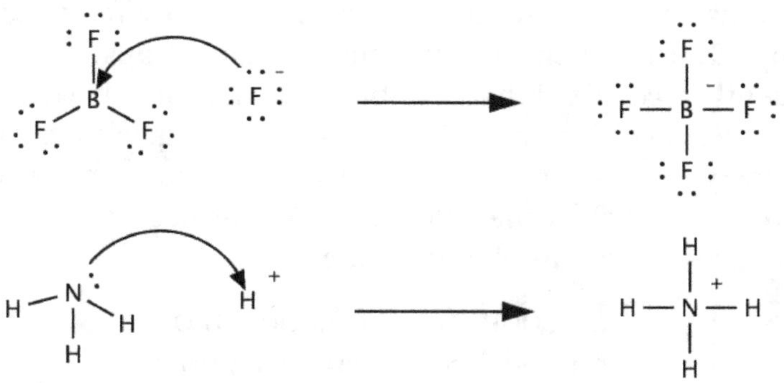

In the first reaction a fluoride ion, F^-, gives up an electron pair to boron trifluoride to form the product tetrafluoroborate. Fluoride "loses" a pair of valence electrons because the electrons shared in the B—F bond are located in the region of space between the two atomic nuclei and are therefore more distant from the fluoride nucleus than they are in the lone fluoride ion. BF_3 is a Lewis acid because it accepts the electron pair from fluoride. This reaction cannot be described in terms of Brønsted theory because there is no proton transfer. The second reaction can be described using either theory. A proton is transferred from an unspecified Brønsted acid to ammonia, a Brønsted base; alternatively, ammonia acts as a Lewis base and transfers a lone pair of electrons to form a bond with a hydrogen ion. The species that gains the electron pair is the Lewis acid; for example, the oxygen atom in H_3O^+ gains a pair of electrons when one of the H—O bonds is broken and the electrons shared in the bond become localized on oxygen. Depending on the context, a Lewis acid may also be described as an oxidizer or an electrophile. Organic Brønsted acids, such as acetic, citric, or oxalic acid, are not Lewis acids. They dissociate in water to produce a Lewis acid, H^+, but at the same time also yield an equal amount of a Lewis base (acetate, citrate, or oxalate, respectively, for the acids mentioned). Few, if any, of the acids discussed in the following are Lewis acids.

Dissociation and Equilibrium

Reactions of acids are often generalized in the form $HA \rightleftharpoons H^+ + A^-$, where HA represents the acid and A^- is the conjugate base. This reaction is referred to as protolysis. The protonated form (HA) of an acid is also sometimes referred to as the free acid.

Acid-base conjugate pairs differ by one proton, and can be interconverted by the addition or removal of a proton (protonation and deprotonation, respectively). Note that the acid can be the charged species and the conjugate base can be neutral in which case the generalized reaction scheme could be written as $HA^+ \rightleftharpoons H^+ + A$. In solution there exists an equilibrium between the acid and its conjugate base. The equilibrium constant K is an expression of the equilibrium concentrations of the molecules or the ions in solution. Brackets indicate concentration, such that $[H_2O]$ means *the concentration of* H_2O. The acid dissociation constant K_a is generally used in the context of acid-base reactions. The numerical value of K_a is equal to the product of the concentrations of the products divided by the concentration of the reactants, where the reactant is the acid (HA) and the products are the conjugate base and H^+.

$$K_a = \frac{[H^+][A^-]}{[HA]}$$

The stronger of two acids will have a higher K_a than the weaker acid; the ratio of hydrogen ions to acid will be higher for the stronger acid as the stronger acid has a greater tendency to lose its proton. Because the range of possible values for K_a spans many orders of magnitude, a more manageable constant, pK_a is more frequently used, where pK_a = $-\log_{10} K_a$. Stronger acids have a smaller pK_a than weaker acids. Experimentally determined pK_a at 25°C in aqueous solution are often quoted in textbooks and reference material.

Nomenclature

In the classical naming system, acids are named according to their anions. That ionic suffix is dropped and replaced with a new suffix (and sometimes prefix), according to the table below. For example, HCl has chloride as its anion, so the -ide suffix makes it take the form hydrochloric acid. In the IUPAC naming system, "aqueous" is simply added to the name of the ionic compound. Thus, for hydrogen chloride, the IUPAC name would be aqueous hydrogen chloride. The prefix "hydro-" is added only if the acid is made up of just hydrogen and one other element.

Classical naming system:

Anion prefix	Anion suffix	Acid prefix	Acid suffix	Example
per	ate	per	ic acid	perchloric acid ($HClO_4$)
	ate		ic acid	chloric acid ($HClO_3$)
	ite		ous acid	chlorous acid ($HClO_2$)
hypo	ite	hypo	ous acid	hypochlorous acid (HClO)
	ide	hydro	ic acid	hydrochloric acid (HCl)

Acid Strength

The strength of an acid refers to its ability or tendency to lose a proton. A strong acid is one that completely dissociates in water; in other words, one mole of a strong acid HA dissolves in water yielding one mole of H^+ and one mole of the conjugate base, A^-, and none of the protonated acid HA. In contrast, a weak acid only partially dissociates and at equilibrium both the acid and the conjugate base are in solution. Examples of strong acids are hydrochloric acid (HCl), hydroiodic acid (HI), hydrobromic acid (HBr), perchloric acid ($HClO_4$), nitric acid (HNO_3) and sulfuric acid (H_2SO_4). In water each of these essentially ionizes 100%. The stronger an acid is, the more easily it loses a proton, H^+. Two key factors that contribute to the ease of deprotonation are the polarity of the

H—A bond and the size of atom A, which determines the strength of the H—A bond. Acid strengths are also often discussed in terms of the stability of the conjugate base.

Stronger acids have a larger K_a and a more negative pK_a than weaker acids.

Sulfonic acids, which are organic oxyacids, are a class of strong acids. A common example is toluenesulfonic acid (tosylic acid). Unlike sulfuric acid itself, sulfonic acids can be solids. In fact, polystyrene functionalized into polystyrene sulfonate is a solid strongly acidic plastic that is filterable.

Superacids are acids stronger than 100% sulfuric acid. Examples of superacids are fluoroantimonic acid, magic acid and perchloric acid. Superacids can permanently protonate water to give ionic, crystalline hydronium "salts". They can also quantitatively stabilize carbocations.

While K_a measures the strength of an acid compound, the strength of an aqueous acid solution is measured by pH, which is an indication of the concentration of hydronium in the solution. The pH of a simple solution of an acid compound in water is determined by the dilution of the compound and the compound's K_a.

Chemical Characteristics

Monoprotic Acids

Monoprotic acids are those acids that are able to donate one proton per molecule during the process of dissociation (sometimes called ionization) as shown below (symbolized by HA):

$$HA_{(aq)} + H_2O_{(l)} \rightleftharpoons H_3O^+_{(aq)} + A^-_{(aq)} \qquad K_a$$

Common examples of monoprotic acids in mineral acids include hydrochloric acid (HCl) and nitric acid (HNO_3). On the other hand, for organic acids the term mainly indicates the presence of one carboxylic acid group and sometimes these acids are known as monocarboxylic acid. Examples in organic acids include formic acid (HCOOH), acetic acid (CH_3COOH) and benzoic acid (C_6H_5COOH).

Polyprotic Acids

Polyprotic acids, also known as polybasic acids, are able to donate more than one proton per acid molecule, in contrast to monoprotic acids that only donate one proton per molecule. Specific types of polyprotic acids have more specific names, such as diprotic acid (two potential protons to donate) and triprotic acid (three potential protons to donate).

A diprotic acid (here symbolized by H_2A) can undergo one or two dissociations depending on the pH. Each dissociation has its own dissociation constant, K_{a1} and K_{a2}.

$$H_2A_{(aq)} + H_2O_{(l)} \rightleftharpoons H_3O^+_{(aq)} + HA^-_{(aq)} \qquad K_{a1}$$
$$HA^-_{(aq)} + H_2O_{(l)} \rightleftharpoons H_3O^+_{(aq)} + A^{2-}_{(aq)} \qquad K_{a2}$$

The first dissociation constant is typically greater than the second; i.e., $K_{a1} > K_{a2}$. For example, sulfuric acid (H_2SO_4) can donate one proton to form the bisulfate anion (HSO_4^-), for which K_{a1} is very large; then it can donate a second proton to form the sulfate anion (SO_4^{2-}), wherein the K_{a2} is intermediate strength. The large K_{a1} for the first dissociation makes sulfuric a strong acid. In a similar manner, the weak unstable carbonic acid (H_2CO_3) can lose one proton to form bicarbonate anion (HCO_3^-) and lose a second to form carbonate anion (CO_3^{2-}). Both K_a values are small, but $K_{a1} > K_{a2}$.

A triprotic acid (H_3A) can undergo one, two, or three dissociations and has three dissociation constants, where $K_{a1} > K_{a2} > K_{a3}$.

$$H_3A_{(aq)} + H_2O_{(l)} \rightleftharpoons H_3O^+_{(aq)} + H_2A^-_{(aq)} \qquad K_{a1}$$
$$H_2A^-_{(aq)} + H_2O_{(l)} \rightleftharpoons H_3O^+_{(aq)} + HA^{2-}_{(aq)} \qquad K_{a2}$$
$$HA^{2-}_{(aq)} + H_2O_{(l)} \rightleftharpoons H_3O^+_{(aq)} + A^{3-}_{(aq)} \qquad K_{a3}$$

An inorganic example of a triprotic acid is orthophosphoric acid (H_3PO_4), usually just called phosphoric acid. All three protons can be successively lost to yield $H_2PO_4^-$, then HPO_4^{2-}, and finally PO_4^{3-}, the orthophosphate ion, usually just called phosphate. Even though the positions of the three protons on the original phosphoric acid molecule are equivalent, the successive K_a values differ since it is energetically less favorable to lose a proton if the conjugate base is more negatively charged. An organic example of a triprotic acid is citric acid, which can successively lose three protons to finally form the citrate ion.

Although the subsequent loss of each hydrogen ion is less favorable, all of the conjugate bases are present in solution. The fractional concentration, α (alpha), for each species can be calculated. For example, a generic diprotic acid will generate 3 species in solution: H_2A, HA^-, and A^{2-}. The fractional concentrations can be calculated as below when given either the pH (which can be converted to the $[H^+]$) or the concentrations of the acid with all its conjugate bases:

$$\alpha_{H2A} = \frac{[H+]^2}{[H+]^2 + [H+]K_1 + K_1K_2} = \frac{[H2A]}{[H2A] + [HA^-] + [A^{2-}]}$$

$$\alpha_{HA^-} = \frac{[H+]K_1}{[H+]^2 + [H+]K_1 + K_1K_2} = \frac{[HA^-]}{[H2A] + [HA^-] + [A^{2-}]}$$

$$\alpha_{A^{2-}} = \frac{K_1K_2}{[H+]^2 + [H+]K_1 + K_1K_2} = \frac{[A^{2-}]}{[H2A] + [HA^-] + [A^{2-}]}$$

A plot of these fractional concentrations against pH, for given K_1 and K_2, is known as a Bjerrum plot. A pattern is observed in the above equations and can be expanded to the general n -protic acid that has been deprotonated i -times:

$$\alpha_{H_{n-i}A^{i-}} = \frac{[H^+]^{n-i}\prod_{j=0}^{i} K_j}{\sum_{i=0}^{n}\left[[H^+]^{n-i}\prod_{j=0}^{i} K_j\right]}$$

where $K_0 = 1$ and the other K-terms are the dissociation constants for the acid.

Neutralization

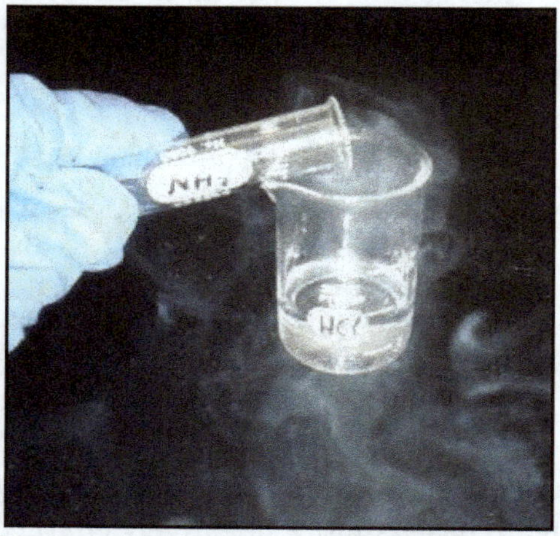

Hydrochloric acid (in beaker) reacting with ammonia fumes to produce ammonium chloride (white smoke).

Neutralization is the reaction between an acid and a base, producing a salt and neutralized base; for example, hydrochloric acid and sodium hydroxide form sodium chloride and water:

$$HCl_{(aq)} + NaOH_{(aq)} \rightarrow H_2O_{(l)} + NaCl_{(aq)}$$

Neutralization is the basis of titration, where a pH indicator shows equivalence point when the equivalent number of moles of a base have been added to an acid. It is often wrongly assumed that neutralization should result in a solution with pH 7.0, which is only the case with similar acid and base strengths during a reaction.

Neutralization with a base weaker than the acid results in a weakly acidic salt. An example is the weakly acidic ammonium chloride, which is produced from the strong acid hydrogen chloride and the weak base ammonia. Conversely, neutralizing a weak acid with a strong base gives a weakly basic salt, e.g. sodium fluoride from hydrogen fluoride and sodium hydroxide.

Weak Acid–weak Base Equilibrium

In order for a protonated acid to lose a proton, the pH of the system must rise above the pK_a of the acid. The decreased concentration of H^+ in that basic solution shifts the equilibrium towards the conjugate base form (the deprotonated form of the acid). In lower-pH (more acidic) solutions, there is a high enough H^+ concentration in the solution to cause the acid to remain in its protonated form.

Solutions of weak acids and salts of their conjugate bases form buffer solutions.

Applications of Acids

There are numerous uses for acids. Acids are often used to remove rust and other corrosion from metals in a process known as pickling. They may be used as an electrolyte in a wet cell battery, such as sulfuric acid in a car battery.

Strong acids, sulfuric acid in particular, are widely used in mineral processing. For example, phosphate minerals react with sulfuric acid to produce phosphoric acid for the production of phosphate fertilizers, and zinc is produced by dissolving zinc oxide into sulfuric acid, purifying the solution and electrowinning.

In the chemical industry, acids react in neutralization reactions to produce salts. For example, nitric acid reacts with ammonia to produce ammonium nitrate, a fertilizer. Additionally, carboxylic acids can be esterified with alcohols, to produce esters.

Acids are used as additives to drinks and foods, as they alter their taste and serve as preservatives. Phosphoric acid, for example, is a component of cola drinks. Acetic acid is used in day-to-day life as vinegar. Carbonic acid is an important part of some cola drinks and soda. Citric acid is used as a preservative in sauces and pickles.

Tartaric acid is an important component of some commonly used foods like unripened mangoes and tamarind. Natural fruits and vegetables also contain acids. Citric acid is present in oranges, lemon and other citrus fruits. Oxalic acid is present in tomatoes, spinach, and especially in carambola and rhubarb; rhubarb leaves and unripe carambolas are toxic because of high concentrations of oxalic acid.

Ascorbic acid (Vitamin C) is an essential vitamin for the human body and is present in such foods as amla (Indian gooseberry), lemon, citrus fruits, and guava.

Certain acids are used as drugs. Acetylsalicylic acid (Aspirin) is used as a pain killer and for bringing down fevers.

Acids play important roles in the human body. The hydrochloric acid present in the stomach aids in digestion by breaking down large and complex food molecules. Amino acids are required for synthesis of proteins required for growth and repair of body tissues. Fatty acids are also required for growth and repair of body tissues. Nucleic acids are important for the manufacturing of DNA and RNA and transmitting of traits to offspring through genes. Carbonic acid is important for maintenance of pH equilibrium in the body.

Acid Catalysis

Acids are used as catalysts in industrial and organic chemistry; for example, sulfuric acid is used in very large quantities in the alkylation process to produce gasoline. Strong acids, such as sulfuric, phosphoric and hydrochloric acids also effect dehydration and condensation reactions. In biochemistry, many enzymes employ acid catalysis.

Biological Occurrence

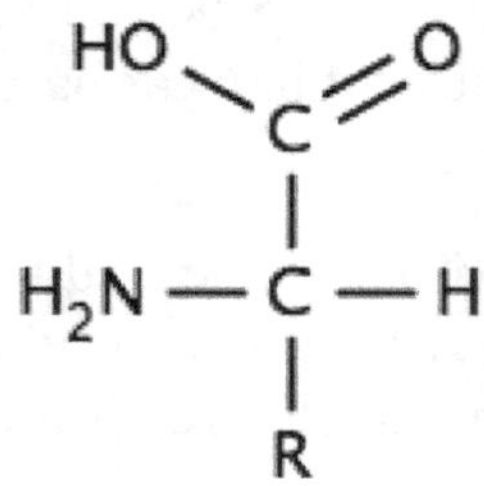

Basic structure of an amino acid.

Many biologically important molecules are acids. Nucleic acids, which contain acidic phosphate groups, include DNA and RNA. Nucleic acids contain the genetic code that determines many of an organism's characteristics, and is passed from parents to offspring. DNA contains the chemical blueprint for the synthesis of proteins which are made up of amino acid subunits. Cell membranes contain fatty acid esters such as phospholipids.

An α-amino acid has a central carbon (the α or *alpha* carbon) which is covalently bonded to a carboxyl group (thus they are carboxylic acids), an amino group, a hydrogen atom and a variable group. The variable group, also called the R group or side chain, determines the identity and many of the properties of a specific amino acid. In glycine, the simplest amino acid, the R group is a hydrogen atom, but in all other amino acids it is contains one or more carbon atoms bonded to hydrogens, and may contain other elements such as sulfur, oxygen or nitrogen. With the exception of glycine, naturally occurring amino acids are chiral and almost invariably occur in the L-configuration. Peptidoglycan, found in some bacterial cell walls contains some D-amino acids. At physiological pH, typically around 7, free amino acids exist in a charged form, where the acidic carboxyl group (-COOH) loses a proton ($-COO^-$) and the basic amine group ($-NH_2$) gains a proton ($-NH_3^+$). The entire molecule has a net neutral charge and is a zwitterion, with the exception of amino acids with basic or acidic side chains. Aspartic acid, for example, possesses one protonated amine and two deprotonated carboxyl groups, for a net charge of −1 at physiological pH.

Fatty acids and fatty acid derivatives are another group of carboxylic acids that play a significant role in biology. These contain long hydrocarbon chains and a carboxylic acid group on one end. The cell membrane of nearly all organisms is primarily made up of a phospholipid bilayer, a micelle of hydrophobic fatty acid esters with polar, hydrophilic phosphate "head" groups. Membranes contain additional components, some of which can participate in acid-base reactions.

In humans and many other animals, hydrochloric acid is a part of the gastric acid secreted within the stomach to help hydrolyze proteins and polysaccharides, as well as converting the inactive pro-enzyme, pepsinogen into the enzyme, pepsin. Some organisms produce acids for defense; for example, ants produce formic acid.

Acid-base equilibrium plays a critical role in regulating mammalian breathing. Oxygen gas (O_2) drives cellular respiration, the process by which animals release the chemical potential energy stored in food, producing carbon dioxide (CO_2) as a byproduct. Oxygen and carbon dioxide are exchanged in the lungs, and the body responds to changing energy demands by adjusting the rate of ventilation. For example, during periods of exertion the body rapidly breaks down stored carbohydrates and fat, releasing CO_2 into the blood stream. In aqueous solutions such as blood CO_2 exists in equilibrium with carbonic acid and bicarbonate ion.

$$CO_2 + H_2O \rightleftharpoons H_2CO_3 \rightleftharpoons H^+ + HCO_3^-$$

It is the decrease in pH that signals the brain to breathe faster and deeper, expelling the excess CO_2 and resupplying the cells with O_2.

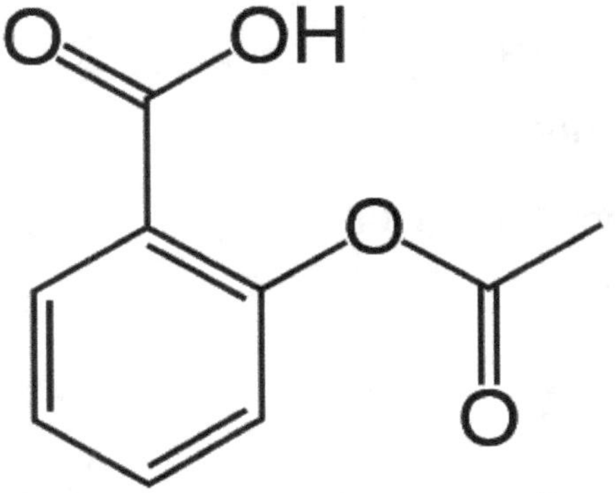

Aspirin (acetylsalicylic acid) is a carboxylic acid.

Cell membranes are generally impermeable to charged or large, polar molecules because of the lipophilic fatty acyl chains comprising their interior. Many biologically important molecules, including a number of pharmaceutical agents, are organic weak acids which can cross the membrane in their protonated, uncharged form but not in their charged form (i.e. as the conjugate base). For this reason the activity of many drugs can be enhanced or inhibited by the use of antacids or acidic foods. The charged form, however, is often more soluble in blood and cytosol, both aqueous environments.

When the extracellular environment is more acidic than the neutral pH within the cell, certain acids will exist in their neutral form and will be membrane soluble, allowing them to cross the phospholipid bilayer. Acids that lose a proton at the intracellular pH will exist in their soluble, charged form and are thus able to diffuse through the cytosol to their target. Ibuprofen, aspirin and penicillin are examples of drugs that are weak acids.

Common Acids

Mineral Acids (Inorganic Acids)

- Hydrogen halides and their solutions: hydrofluoric acid (HF), hydrochloric acid (HCl), hydrobromic acid (HBr), hydroiodic acid (HI)
- Halogen oxoacids: hypochlorous acid (HClO), chlorous acid ($HClO_2$), chloric acid ($HClO_3$), perchloric acid ($HClO_4$), and corresponding analogs for bromine and iodine
    - Hypofluorous acid (HFO), the only known oxoacid for fluorine.
- Sulfuric acid (H_2SO_4)
- Fluorosulfuric acid (HSO_3F)
- Nitric acid (HNO_3)
- Phosphoric acid (H_3PO_4)
- Fluoroantimonic acid ($HSbF_6$)
- Fluoroboric acid (HBF_4)
- Hexafluorophosphoric acid (HPF_6)
- Chromic acid (H_2CrO_4)
- Boric acid (H_3BO_3)

Sulfonic Acids

A sulfonic acid has the general formula $RS(=O)_2{-}OH$, where R is an organic radical.

- Methanesulfonic acid (or mesylic acid, CH_3SO_3H)
- Ethanesulfonic acid (or esylic acid, $CH_3CH_2SO_3H$)
- Benzenesulfonic acid (or besylic acid, $C_6H_5SO_3H$)
- p-Toluenesulfonic acid (or tosylic acid, $CH_3C_6H_4SO_3H$)

- Trifluoromethanesulfonic acid (or triflic acid, CF_3SO_3H)
- Polystyrene sulfonic acid (sulfonated polystyrene, $[CH_2CH(C_6H_4)SO_3H]_n$)

Carboxylic Acids

A carboxylic acid has the general formula R-C(O)OH, where R is an organic radical. The carboxyl group -C(O)OH contains a carbonyl group, C=O, and a hydroxyl group, O-H.

- Acetic acid (CH_3COOH)
- Citric acid ($C_6H_8O_7$)
- Formic acid (HCOOH)
- Gluconic acid $HOCH_2$-$(CHOH)_4$-COOH
- Lactic acid (CH_3-CHOH-COOH)
- Oxalic acid (HOOC-COOH)
- Tartaric acid (HOOC-CHOH-CHOH-COOH)

Halogenated Carboxylic Acids

Halogenation at alpha position increases acid strength, so that the following acids are all stronger than acetic acid.

- Fluoroacetic acid
- Trifluoroacetic acid
- Chloroacetic acid
- Dichloroacetic acid
- Trichloroacetic acid

Vinylogous Carboxylic Acids

Normal carboxylic acids are the direct union of a carbonyl group and a hydroxyl group. In vinylogous carboxylic acids, a carbon-carbon double bond separates the carbonyl and hydroxyl groups.

- Ascorbic acid

Nucleic Acids

- Deoxyribonucleic acid (DNA)
- Ribonucleic acid (RNA)

Base

In chemistry, bases are substances that, in aqueous solution, are slippery to the touch, taste astringent, change the color of indicators (e.g., turn red litmus paper blue), react with acids to form salts, promote certain chemical reactions (base catalysis), accept protons from any proton donor, and/or contain completely or partially displaceable OH^- ions. Examples of bases are the hydroxides of the alkali metals and alkaline earth metals (NaOH, $Ca(OH)_2$, etc.).

Soaps are weak bases formed by the reaction of fatty acids with sodium hydroxide or potassium hydroxide.

These particular substances produce hydroxide ions (OH^-) in aqueous solutions, and are thus classified as Arrhenius bases. For a substance to be classified as an Arrhenius base, it must produce hydroxide ions in an aqueous solution. In order to do so, Arrhenius believed the base must contain hydroxide in the formula. This makes the Arrhenius model limited, as it cannot explain the basic properties of aqueous solutions of ammonia (NH_3) or its organic derivatives (amines). There are also bases that do not contain a hydroxide ion but nevertheless react with water, resulting in an increase in the concentration of the hydroxide ion. An example of this is the reaction between ammonia and water to produce ammonium and hydroxide. In this reaction ammonia is the base because it accepts a proton from the water molecule. Ammonia and other bases similar to it usually have the ability to form a bond with a proton due to the unshared pair of electrons that they possess. In the more general Brønsted–Lowry acid–base theory, a base is a substance that can accept hydrogen cations (H^+)—otherwise known as protons. In the Lewis model, a base is an electron pair donor.

In water, by altering the autoionization equilibrium, bases yield solutions in which the hydrogen ion activity is lower than it is in pure water, i.e., the water has a pH higher than 7.0 at standard conditions. A soluble base is called an alkali if it contains and releases OH^- ions quantitatively. However, it is important to realize that basicity is not the same as alkalinity. Metal oxides, hydroxides, and especially alkoxides are basic, and counteranions of weak acids are weak bases.

Bases can be thought of as the chemical opposite of acids. However, some strong acids are able to act as bases. Bases and acids are seen as opposites because the effect of an acid is to increase the hydronium (H_3O^+) concentration in water, whereas bases reduce this concentration. A reaction between an acid and base is called neutralization. In a neutralization reaction, an aqueous solution of a base reacts with an aqueous solution of an acid to produce a solution of water and salt in which the salt separates into its component ions. If the aqueous solution is saturated with a given salt solute, any additional such salt precipitates out of the solution.

The notion of a base as a concept in chemistry was first introduced by the French chemist Guillaume François Rouelle in 1754. He noted that acids, which at that time were mostly volatile liquids (like acetic acid), turned into solid salts only when combined with specific substances. Rouelle considered that such a substance serves as a "base" for the salt, giving the salt a "concrete or solid form".

Properties

General properties of bases include:

- Concentrated or strong bases are caustic on organic matter and react violently with acidic substances.
- Aqueous solutions or molten bases dissociate in ions and conduct electricity.
- Reactions with indicators: bases turn red litmus paper blue, phenolphthalein pink, keep bromothymol blue in its natural colour of blue, and turn methyl orange yellow.
- The pH of a basic solution at standard conditions is greater than seven.
- Bases are bitter in taste.

Reactions Between Bases and Water

The following reaction represents the general reaction between a base (B) and water to produce a conjugate acid (BH^+) and a conjugate base (OH^-):

$$B_{(aq)} + H_2O_{(l)} \rightleftharpoons BH^+_{(aq)} + OH^-_{(aq)}$$

The equilibrium constant, K_b, for this reaction can be found using the following general equation:

$$K_b = [BH^+][OH^-]/[B]$$

In this equation, both the base (B) and the extremely strong base (the conjugate base) compete with one another for the proton. As a result, bases that react with water have relatively small equilibrium constant values. The base is weaker when it has a lower equilibrium constant value.

Neutralization of Acids

Bases react with acids to neutralize each other at a fast rate both in water and in alcohol. When dissolved in water, the strong base sodium hydroxide ionizes into hydroxide and sodium ions:

$$NaOH \rightarrow Na^{+} + OH^{-}$$

and similarly, in water the acid hydrogen chloride forms hydronium and chloride ions:

$$HCl + H_2O \rightarrow H_3O^{+} + Cl^{-}$$

When the two solutions are mixed, the H_3O^{+} and OH^{-} ions combine to form water molecules:

$$H_3O^{+} + OH^{-} \rightarrow 2\ H_2O$$

If equal quantities of NaOH and HCl are dissolved, the base and the acid neutralize exactly, leaving only NaCl, effectively table salt, in solution.

Weak bases, such as baking soda or egg white, should be used to neutralize any acid spills. Neutralizing acid spills with strong bases, such as sodium hydroxide or potassium hydroxide can cause a violent exothermic reaction, and the base itself can cause just as much damage as the original acid spill.

Alkalinity of Non-hydroxides

Bases are generally compounds that can neutralize an amount of acids. Both sodium carbonate and ammonia are bases, although neither of these substances contains OH− groups. Both compounds accept H^{+} when dissolved in protic solvents such as water:

$$Na_2CO_3 + H_2O \rightarrow 2\ Na^{+} + HCO_3^{-} + OH^{-}$$

$$NH_3 + H_2O \rightarrow NH_4^{+} + OH^{-}$$

From this, a pH, or acidity, can be calculated for aqueous solutions of bases. Bases also directly act as electron-pair donors themselves:

$$CO_3^{2-} + H^{+} \rightarrow HCO_3^{-}NH_3 + H^{+} \rightarrow NH_4^{+}$$

A base is also defined as a molecule that has the ability to accept an electron pair bond by entering another atom's valence shell through its possession of one electron pair. There are a limited number of elements that have atoms with the ability to provide a molecule with basic properties. Carbon can act as a base as well as nitrogen and oxygen. Fluorine and sometimes rare gases possess this ability as well. This occurs typically in compounds such as butyl lithium, alkoxides, and metal amides such as sodium amide. Bases of carbon, nitrogen and oxygen without resonance stabilization are usually very strong, or superbases, which cannot exist in a water solution due to the acidity of water. Resonance stabilization, however, enables weaker bases such as carboxylates; for example, sodium acetate is a weak base.

Strong Bases

A strong base is a basic chemical compound that can remove a proton (H+) from (or *deprotonate*) a molecule of a very weak acid in an acid-base reaction. Common examples of strong bases include hydroxides of alkali metals and alkaline earth metals like NaOH and Ca(OH) 2 respectively. Due to their low solubility, some bases, such as alkaline earth hydroxides, can be used when the solubility factor is not taken into account. One advantage of this low solubility is that "many antacids were suspensions of metal hydroxides such as aluminum hydroxide and magnesium hydroxide." These compounds have low solubility and have the ability to stop an increase in the concentration of the hydroxide ion, preventing the harm of the tissues in the mouth, esophagus, and stomach. As the reaction continues and the salts dissolve, the stomach acid reacts with the hydroxide produced by the suspensions. Very strong bases hydrolyze in water almost completely, resulting in the "leveling effect." In this process, the water molecule combines with a strong base, due to the water's amphoteric ability; and, a hydroxide ion is released. Very strong bases can even deprotonate very weakly acidic C–H groups in the absence of water. Here is a list of several strong bases:

- Lithium hydroxide (LiOH)
- Sodium hydroxide (NaOH)
- Potassium hydroxide (KOH)
- Rubidium hydroxide (RbOH)
- Cesium hydroxide (CsOH)
- Magnesium hydroxide (Mg(OH)2) (in solution; but the solid has only limited solubility and thus is weakly basic)
- Calcium hydroxide (Ca(OH)2)
- Strontium hydroxide (Sr(OH)2)
- Barium hydroxide (Ba(OH)2)

The cations of these strong bases appear in the first and second groups of the periodic table (alkali and earth alkali metals).

Acids with a pK_a of more than about 13 are considered very weak, and their conjugate bases are strong bases.

Strong bases are able to form stable compounds by combining with a strong acid. Weak bases, however, do not possess the ability to form stable compounds by combining with a weak acid.

Superbases

Group 1 salts of carbanions, amides, and hydrides tend to be even stronger bases due to the extreme weakness of their conjugate acids, which are stable hydrocarbons, amines, and dihydrogen. Usually these bases are created by adding pure alkali metals such as sodium into the conjugate acid. They are called *superbases*, and it is impossible to keep them in water solution because they are stronger bases than the hydroxide ion. As such, they deprotonate the conjugate acid water. For example, the ethoxide ion (conjugate base of ethanol) in the presence of water undergoes this reaction.

$$CH_3CH2O^- + H_2O \rightarrow CH_3CH_2OH + OH^-$$

Examples of superbases are:

- Butyl lithium (n-C_4H_9Li)
- Lithium diisopropylamide (LDA) $[(CH_3)_2CH]_2NLi$
- Lithium diethylamide (LDEA) $(C_2H_5)_2NLi$
- Sodium amide ($NaNH_2$)
- Sodium hydride (NaH)
- Lithium bis(trimethylsilyl)amide $[(CH_3)_3Si_2NLi$

Neutral Bases

When a neutral base forms a bond with a neutral acid, a condition of electric stress occurs. The acid and the base share the electron pair that formerly only belonged to the base. As a result, a high dipole moment is created, which can only be destroyed by rearranging the molecules.

Bases as Catalysts

Basic substances can be used as insoluble heterogeneous catalysts for chemical reactions. Some examples are metal oxides such as magnesium oxide, calcium oxide, and barium oxide as well as potassium fluoride on alumina and some zeolites. Many transition metals make good catalysts, many of which form basic substances. Basic catalysts have been used for hydrogenations, the migration of double bonds, in the Meerwein-Ponndorf-Verley reduction, the Michael reaction, and many other reactions. Both CaO and BaO can be highly active catalysts if they are treated with high temperature heat.

Solid Bases

Examples of solid bases include:

- Oxide mixtures: SiO_2, Al_2O_3; MgO, SiO_2; CaO, SiO_2
- Mounted bases: $LiCO_3$ on silica; NR_3, NH_3, KNH_2 on alumina; NaOH, KOH mounted on silica on alumina
- Inorganic chemicals: BaO, $KNaCO_3$, BeO, MgO, CaO, KCN
- Anion exchange resins
- Charcoal that has been treated at 900 degrees Celsius or activates with N_2O, NH_3, $ZnCl_2$-NH_4Cl-CO_2

Depending on a solid surface's ability to successfully form a conjugate base by absorbing an electrically neutral acid, the basic strength of the surface is determined. "The number of basic sites per unit surface area of the solid" is used to express how much base is found on a solid base catalyst. Scientists have developed two methods to measure the amount of basic sites: titration with benzoic acid using indicators and gaseous acid adsorption. A solid with enough basic strength will absorb an electrically neutral acid indicator and cause the acid indicator's color to change to the color of its conjugate base. When performing the gaseous acid adsorption method, nitric oxide is used. The basic sites are then determined using the amount of carbon dioxide than is absorbed.

Weak Bases

When there is a hydrogen ion gradient between two sides of the biological membrane, the concentration of some weak bases are focused on only one side of the membrane. Weak bases tend to build up in acidic fluids. Acid gastric contains a higher concentration of weak base than plasma. Acid urine, compared to alkaline urine, excretes weak bases at a faster rate.

Uses of Bases

- Sodium hydroxide is used in manufacture of soap, paper and a synthetic fiber called "rayon".
- Calcium hydroxide (slaked lime) is used in the manufacture of bleaching powder.
- Calcium hydroxide is also used to clean the sulfur dioxide, which is caused by exhaust, that is found in power plants and factories.
- Magnesium hydroxide is used as an 'antacid' to neutralize excess acid in the stomach and cure indigestion.

- Sodium carbonate is used as washing soda and for softening hard water.
- Sodium hydrogen carbonate is used as baking soda in cooking food, for making baking powders, as an antacid to cure indigestion and in soda acid fire extinguisher.
- Ammonium hydroxide is used to remove grease stains from clothes.

Acidity of Bases

The number of ionizable hydroxide (OH-) ions present in one molecule of base is called the acidity of bases. On the basis of acidity bases can be classified into three types: monoacidic, diacidic and triacidic.

Monoacidic Bases

Sodium hydroxide.

When one molecule of a base via complete ionization produces one hydroxide ion, the base is said to be a monoacidic base. Examples of monoacidic bases are:

Sodium hydroxide, potassium hydroxide, silver hydroxide, ammonium hydroxide, etc.

Diacidic Bases

Barium hydroxide.

When one molecule of base via complete ionization produces two hydroxide ions, the base is said to be diacidic. Examples of diacidic bases are:

Barium hydroxide, magnesium hydroxide, calcium hydroxide, zinc hydroxide, iron (II) hydroxide, tin (II) hydroxide, lead (II) hydroxide, copper (II) hydroxide, etc.

Triacidic Bases

When one molecule of base via complete ionization produces three hydroxide ions, the base is said to be triacidic. Examples of triacidic bases are:

Aluminium hydroxide, ferrous hydroxide, Gold Trihydroxide.

Etymology of the Term

The concept of base stems from an older alchemichal notion of "the matrix":

The term "base" appears to have been first used in 1717 by the French chemist, Louis Lémery, as a synonym for the older Paracelsian term "matrix". In keeping with 16th-century animism, Paracelsus had postulated that naturally occurring salts grew within the earth as a result of a universal acid or seminal principle having impregnated an earthy matrix or womb... Its modern meaning and general introduction into the chemical vocabulary, however, is usually attributed to the French chemist, Guillaume-François Rouelle... Rouelle explicitly defined a neutral salt as the product formed by the union of an acid with any substance, be it a water-soluble alkali, a volatile alkali, an absorbent earth, a metal, or an oil, capable of serving as "a base" for the salt "by giving it a concrete or solid form." Most acids known in the 18th century were volatile liquids or "spirits" capable of distillation, whereas salts, by their very nature, were crystalline solids. Hence it was the substance that neutralized the acid which supposedly destroyed the volatility or spirit of the acid and which imparted the property of solidity (i.e., gave a concrete base) to the resulting salt.

— *William Jensen, The origin of the term "base"*

Acid–base Reaction

An acid–base reaction is a chemical reaction that occurs between an acid and a base. Several theoretical frameworks provide alternative conceptions of the reaction mechanisms and their application in solving related problems; these are called acid–base theories, for example, Brønsted–Lowry acid–base theory. Their importance becomes apparent in analyzing acid–base reactions for gaseous or liquid species, or when acid or base character may be somewhat less apparent. The first of these concepts was provided by the French chemist Antoine Lavoisier, around 1776.

Acid–base Definitions

Lavoisier's Oxygen Theory of Acids

The first scientific concept of acids and bases was provided by Lavoisier in around 1776. Since Lavoisier's knowledge of strong acids was mainly restricted to oxoacids, such as HNO_3 (nitric acid) and H_2SO_4 (sulfuric acid), which tend to contain central atoms in high oxidation states surrounded by oxygen, and since he was not aware of the true composition of the hydrohalic acids (HF, HCl, HBr, and HI), he defined acids in terms of their containing *oxygen*, which in fact he named from Greek words meaning "acid-former" meaning "acid" or "sharp" and (*geinomai*) meaning "engender"). The Lavoisier definition was held as absolute truth for over 30 years, until the 1810 article and subsequent section by Sir Humphry Davy in which he proved the lack of oxygen in H_2S, H_2Te, and the hydrohalic acids. However, Davy failed to develop a new theory, concluding that "acidity does not depend upon any particular elementary substance, but upon peculiar arrangement of various substances". One notable modification of oxygen theory was provided by Berzelius, who stated that acids are oxides of nonmetals while bases are oxides of metals.

Liebig's Hydrogen Theory of Acids

In 1838, Justus von Liebig proposed that an acid is a hydrogen-containing substance in which the hydrogen could be replaced by a metal. This redefinition was based on his extensive work on the chemical composition of organic acids, finishing the doctrinal shift from oxygen-based acids to hydrogen-based acids started by Davy. Liebig's definition, while completely empirical, remained in use for almost 50 years until the adoption of the Arrhenius definition.

Arrhenius Definition

Svante Arrhenius.

The first modern definition of acids and bases was devised by Svante Arrhenius. A hydrogen theory of acids, it followed from his 1884 work with Friedrich Wilhelm Ostwald in establishing the presence of ions in aqueous solution and led to Arrhenius receiving the Nobel Prize in Chemistry in 1903.

As defined by Arrhenius:

- *an Arrhenius acid* is a substance that dissociates in water to form hydrogen ions (H^+); that is, an acid increases the concentration of H^+ ions in an aqueous solution.

This causes the protonation of water, or the creation of the hydronium (H_3O^+) ion. Thus, in modern times, the symbol H^+ is interpreted as a shorthand for H_3O^+, because it is now known that a bare proton does not exist as a free species in aqueous solution.

- *an Arrhenius base* is a substance that dissociates in water to form hydroxide (OH^-) ions; that is, a base increases the concentration of OH^- ions in an aqueous solution.

The Arrhenius definitions of acidity and alkalinity are restricted to aqueous solutions, and refer to the concentration of the solvent ions. Under this definition, pure H_2SO_4 and HCl dissolved in toluene are not acidic, and molten NaOH and solutions of calcium amide in liquid ammonia are not alkaline.

Overall, to qualify as an Arrhenius acid, upon the introduction to water, the chemical must either cause, directly or otherwise:

- an increase in the aqueous hydronium concentration, or
- a decrease in the aqueous hydroxide concentration.

Conversely, to qualify as an Arrhenius base, upon the introduction to water, the chemical must either cause, directly or otherwise:

- a decrease in the aqueous hydronium concentration, or
- an increase in the aqueous hydroxide concentration.

The reaction of an acid with a base is called a neutralization reaction. The products of this reaction are a salt and water.

acid + base → salt + water

In this traditional representation an acid–base neutralization reaction is formulated as a double-replacement reaction. For example, the reaction of hydrochloric acid, HCl, with sodium hydroxide, NaOH, solutions produces a solution of sodium chloride, NaCl, and some additional water molecules.

$$HCl(aq) + NaOH(aq) \rightarrow NaCl(aq) + H_2O$$

The modifier (aq) in this equation is important. It was implied by Arrhenius, not included explicitly. It indicates that the substances are dissolved in water. In fact though all three substances, HCl, NaOH and NaCl are capable of existing as pure compounds, in aqueous solutions they are fully dissociated into the (aquated) ions H^+, Cl^-, Na^+ and OH^-.

Brønsted–Lowry Definition

Johannes Nicolaus Brønsted and Thomas Martin Lowry.

The Brønsted–Lowry definition, formulated in 1923, independently by Johannes Nicolaus Brønsted in Denmark and Martin Lowry in England, is based upon the idea of protonation of bases through the de-protonation of acids – that is, the ability of acids to "donate" hydrogen ions (H^+)—otherwise known as protons—to bases, which "accept" them.

An acid–base reaction is, thus, the removal of a hydrogen ion from the acid and its addition to the base. The removal of a hydrogen ion from an acid produces its *conjugate base*, which is the acid with a hydrogen ion removed. The reception of a proton by a base produces its *conjugate acid*, which is the base with a hydrogen ion added.

Unlike the previous definitions, the Brønsted–Lowry definition does not refer to the formation of salt and solvent, but instead to the formation of *conjugate acids* and *conjugate bases*, produced by the transfer of a proton from the acid to the base. In this approach, acids and bases are fundamentally different in behavior from salts, which are seen as electrolytes, subject to the theories of Debye, Onsager, and others. An acid and a base react not to produce a salt and a solvent, but to form a new acid and a new base. The concept of neutralization is thus absent. Brønsted–Lowry acid–base behavior is formally independent of any solvent, making it more all-encompassing than the Arrhenius model.

The general formula for acid–base reactions according to the Brønsted–Lowry definition is:

$$HA + B \rightarrow BH^+ + A^-$$

where HA represents the acid, B represents the base, BH^+ represents the conjugate acid of B, and A^- represents the conjugate base of HA.

For example, a Brønsted-Lowry model for the dissociation of hydrochloric acid (HCl) in aqueous solution would be the following:

$$HCl + H_2O \rightleftharpoons H_3O^+ + Cl^-$$

The removal of H^+ from the HCl produces the chloride ion, Cl^-, the conjugate base of the acid. The addition of H^+ to the H_2O (acting as a base) forms the hydronium ion, H_3O^+, the conjugate acid of the base.

Water is amphoteric—that is, it can act as both an acid and a base. The Brønsted-Lowry model explains this, showing the dissociation of water into low concentrations of hydronium and hydroxide ions:

$$H_2O + H_2O \rightleftharpoons H_3O^+ + OH^-$$

This equation is demonstrated in the image below:

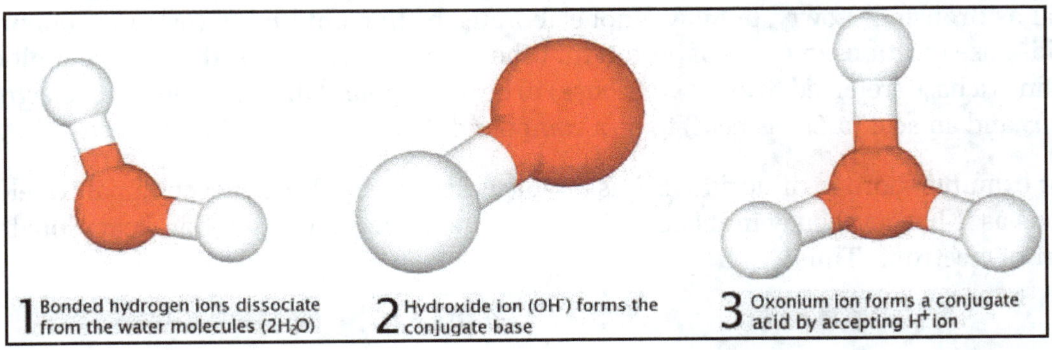

Here, one molecule of water acts as an acid, donating an H^+ and forming the conjugate base, OH^-, and a second molecule of water acts as a base, accepting the H^+ ion and forming the conjugate acid, H_3O^+.

As an example of water acting as an acid, consider an aqueous solution of pyridine, C_5H_5N.

$$C_5H_5N + H_2O \rightleftharpoons [C_5H_5NH]^+ + OH^-$$

In this example, a water molecule is split into a hydrogen ion, which is donated to a pyridine molecule, and an hydroxide ion.

In the Brønsted-Lowry model, the solvent does not necessarily have to be water. For example, consider what happens when acetic acid, CH_3COOH, dissolves in liquid ammonia.

$$CH_3COOH + NH_3 \rightleftharpoons NH_4^+ + CH_3COO^-$$

An H^+ ion is removed from acetic acid, forming its conjugate base, the acetate ion, CH_3COO^-. The addition of an H^+ ion to an ammonia molecule of the solvent creates its conjugate acid, the ammonium ion, NH_4^+.

The Brønsted–Lowry model calls hydrogen-containing substances (like HCl) acids. Thus, some substances, which many chemists considered to be acids, such as SO_3 or BCl_3, are excluded from this classification due to lack of hydrogen. Gilbert N. Lewis wrote in 1938, "To restrict the group of acids to those substances that contain hydrogen interferes as seriously with the systematic understanding of chemistry as would the restriction of the term oxidizing agent to substances containing oxygen." Furthermore, KOH and KNH_2 are not considered Brønsted bases, but rather salts containing the bases OH^- and NH_2^-.

Lewis Definition

The hydrogen requirement of Arrhenius and Brønsted–Lowry was removed by the Lewis definition of acid–base reactions, devised by Gilbert N. Lewis in 1923, in the same year as Brønsted–Lowry, but it was not elaborated by him until 1938. Instead of defining acid–base reactions in terms of protons or other bonded substances, the Lewis definition defines a base (referred to as a *Lewis base*) to be a compound that can donate an *electron pair*, and an acid (a *Lewis acid*) to be a compound that can receive this electron pair.

For example, boron trifluoride, BF_3 is a typical Lewis acid. It can accept a pair of electrons as it has a vacancy in its octet. The fluoride ion has a full octet and can donate a pair of electrons. Thus,

$$BF_3 + F^- \rightarrow BF_4^-$$

is a typical Lewis acid, Lewis base reaction. All compounds of group 13 elements with a formula AX_3 can behave as Lewis acids. Similarly, compounds of group 15 elements with a formula DY_3, such as amines, NR_3, and phosphines, PR_3, can behave as Lewis bases. Adducts between them have the formula $X_3A \leftarrow DY_3$ with a dative covalent bond, shown symbolically as ←, between the atoms A (acceptor) and D (donor). Compounds of group 16 with a formula DX_2 may also act as Lewis bases; in this way, a compound like an ether, R_2O, or a thioether, R_2S, can act as a Lewis base. The Lewis definition is not limited to these examples. For instance, carbon monoxide acts as a Lewis base when it forms an adduct with boron trifluoride, of formula $F_3B \leftarrow CO$.

Adducts involving metal ions are referred to as co-ordination compounds; each ligand donates a pair of electrons to the metal ion. The reaction,

$$\left[Ag(H_2O)_4\right]^+ + 2NH_3 \rightarrow \left[Ag(NH_3)_2\right]^+ + 4H_2O$$

can be seen as an acid–base reaction in which a stronger base (ammonia) replaces a weaker one (water).

The Lewis and Brønsted–Lowry definitions are consistent with each other since the reaction,

$$H^{+} + OH^{-} \rightleftharpoons H_2O$$

is an acid–base reaction in both theories.

Solvent System Definition

One of the limitations of the Arrhenius definition is its reliance on water solutions. Edward Curtis Franklin studied the acid–base reactions in liquid ammonia in 1905 and pointed out the similarities to the water-based Arrhenius theory. Albert F. O. Germann, working with liquid phosgene, COCl2, formulated the solvent-based theory in 1925, thereby generalizing the Arrhenius definition to cover aprotic solvents.

Germann pointed out that in many solutions, there are ions in equilibrium with the neutral solvent molecules:

- solvonium: A generic name for a positive ion.
- solvate: A generic name for a negative ion.

For example, water and ammonia undergo such dissociation into hydronium and hydroxide, and ammonium and amide, respectively:

$$2H_2O \rightleftharpoons H_3O^{+} + OH^{-}$$

$$2NH_3 \rightleftharpoons NH_4^{+} + NH_2^{-}$$

Some aprotic systems also undergo such dissociation, such as dinitrogen tetroxide into nitrosonium and nitrate, antimony trichloride into dichloroantimonium and tetrachloroantimonate, and phosgene into chlorocarboxonium and chloride:

$$N_2O_4 \rightleftharpoons NO^{+} + NO_3^{-}$$

$$2SbCl_3 \rightleftharpoons SbCl_2^{+} + SbCl_4^{-}$$

$$COCl_2 \rightleftharpoons COCl^{+} + Cl^{-}$$

A solute that causes an increase in the concentration of the solvonium ions and a decrease in the concentration of solvate ions is defined as an *acid*. A solute that causes an increase in the concentration of the solvate ions and a decrease in the concentration of the solvonium ions is defined as a *base*.

Thus, in liquid ammonia, KNH_2 (supplying NH_2^-) is a strong base, and NH_4NO_3 (supplying NH_4^+) is a strong acid. In liquid sulfur dioxide (SO_2), thionyl compounds (supplying SO^{2+}) behave as acids, and sulfites (supplying SO_3^{2-}) behave as bases.

The non-aqueous acid–base reactions in liquid ammonia are similar to the reactions in water:

$$\underset{\text{(base)}}{2\,NaNH_2} + \underset{\text{(amphiphilic amide)}}{Zn(NH_2)_2} \rightarrow Na_2[Zn(NH_2)_4]$$

$$\underset{\text{(acid)}}{2\,NH_4I} + \underset{\text{(amphiphilic amide)}}{Zn(NH_2)_2} \rightarrow [Zn(NH)_4)]I_2$$

Nitric acid can be a base in liquid sulfuric acid:

$$\underset{\text{(base)}}{HNO_3} + 2H_2SO_4 \rightarrow NO + 2 + H_3O + + 2HSO_4^-$$

The unique strength of this definition shows in describing the reactions in aprotic solvents; for example, in liquid N_2O_4:

$$\underset{\text{(base)}}{AgNO_3} + \underset{\text{(acid)}}{NOCl} \rightarrow \underset{\text{(solvent)}}{N_2O_4} + \underset{\text{(salt)}}{AgCl}$$

Because the solvent system definition depends on the solute as well as on the solvent itself, a particular solute can be either an acid or a base depending on the choice of the solvent: $HClO_4$ is a strong acid in water, a weak acid in acetic acid, and a weak base in fluorosulfonic acid; this characteristic of the theory has been seen as both a strength and a weakness, because some substances (such as SO_3 and NH_3) have been seen to be acidic or basic on their own right. On the other hand, solvent system theory has been criticized as being too general to be useful. Also, it has been thought that there is something intrinsically acidic about hydrogen compounds, a property not shared by non-hydrogenic solvonium salts.

Lux–Flood Definition

This acid–base theory was a revival of oxygen theory of acids and bases, proposed by German chemist Hermann Lux in 1939, further improved by Håkon Flood circa 1947 and is still used in modern geochemistry and electrochemistry of molten salts. This definition describes an acid as an oxide ion (O^{2-}) acceptor and a base as an oxide ion donor. For example:

$$\underset{\text{(base)}}{MgO} + \underset{\text{(acid)}}{CO_2} \rightarrow MgCO_3$$

$$\underset{\text{(base)}}{CaO} + \underset{\text{(acid)}}{SiO_2} \rightarrow CaSiO_3$$

$$\underset{\text{(base)}}{NO_3^-} + \underset{\text{(acid)}}{S2O_7^{2-}} \rightarrow NO + 2 + 2SO_4^{2-}$$

This theory is also useful in the systematisation of the reactions of noble gas compounds, especially the xenon oxides, fluorides, and oxofluorides.

Usanovich Definition

Mikhail Usanovich developed a general theory that does not restrict acidity to hydrogen-containing compounds, but his approach, published in 1938, was even more general than Lewis theory. Usanovich's theory can be summarized as defining an acid as anything that accepts negative species or donates positive ones, and a base as the reverse. This defined the concept of redox (oxidation-reduction) as a special case of acid–base reactions.

Some examples of Usanovich acid–base reactions include:

$$\underset{(\text{base})}{Na_2O} + \underset{(\text{acid})}{SO_3} \rightarrow 2Na^+ + SO_4^{2-} \left(\text{species exchanged}: O^{2-}\ \text{anion}\right)$$

$$\underset{(\text{base})}{3(NH_4)_2S} + \underset{(\text{acid})}{Sb_2S_5} \rightarrow 6NH_4^+ + 2SbS_4^{3-} \left(\text{species exchanged}: 3S^{2-}\ \text{anions}\right)$$

$$\underset{(\text{base})}{2Na} + \underset{(\text{acid})}{Cl_2} \rightarrow 2Na^+ + 2Cl^- \left(\text{species exchanged}: 2\ \text{electrons}\right)$$

Pearson Definition

In 1963, Ralph Pearson proposed a qualitative concept known as Hard Soft Acid Base principle. later made quantitative with help of Robert Parr in 1984. 'Hard' applies to species that are small, have high charge states, and are weakly polarizable. 'Soft' applies to species that are large, have low charge states and are strongly polarizable. Acids and bases interact, and the most stable interactions are hard–hard and soft–soft. This theory has found use in organic and inorganic chemistry.

Acid–base Equilibrium

The reaction of a strong acid with a strong base is essentially a quantitative reaction. For example,

$$HCl_{(aq)} + Na(OH)_{(aq)} \rightarrow H_2O + NaCl_{(aq)}$$

In this reaction both the sodium and chloride ions are spectators as the neutralization reaction,

$$H^+ + OH^- \rightarrow H_2O$$

does not involve them. With weak bases addition of acid is not quantitative because a solution of a weak base is a buffer solution. A solution of a weak acid is also a buffer solution. When a weak acid reacts with a weak base an equilibrium mixture is produced. For example, adenine, written as AH can react with a hydrogen phosphate ion, HPO_4^{2-}.

$$AH + HPO_4^{2-} \rightleftharpoons A^- + H_2PO_4^-$$

The equilibrium constant for this reaction can be derived from the acid dissociation constants of adenine and of the dihydrogen phosphate ion.

$$\left[A^{-}\right]\left[H^{+}\right] = K_{a1}\left[AH\right]$$

$$\left[HPO_4^{2-}\right]\left[H^{+}\right] = K_{a2}\left[H2PO_4^{-}\right]$$

The notation [X] signifies "concentration of X". When these two equations are combined by eliminating the hydrogen ion concentration, an expression for the equilibrium constant, *K* is obtained.

$$\left[A^{-}\right]\left[H_2PO_4^{-}\right] = K\left[AH\right]\left[HPO_4^{2-}\right]; \quad K = \frac{K_{a1}}{K_{a2}}$$

Acid–Alkali Reaction

An acid–alkali reaction is a special case of an acid–base reaction, where the base used is also an alkali. When an acid reacts with an alkali salt (a metal hydroxide), the product is a metal salt and water. Acid–alkali reactions are also neutralization reactions.

In general, acid–alkali reactions can be simplified to,

$$OH^{-}_{(aq)} + H^{+}_{(aq)} \rightarrow H_2O$$

by omitting spectator ions.

Acids are in general pure substances that contain hydrogen cations (H^{+}) or cause them to be produced in solutions. Hydrochloric acid (HCl) and sulfuric acid (H_2SO_4) are common examples. In water, these break apart into ions:

$$HCl \rightarrow H^{+}_{(aq)} + Cl^{-}_{(aq)}$$

$$H_2SO_4 \rightarrow H^{+}_{(aq)} + HSO^{-}_{4(aq)}$$

The alkali breaks apart in water, yielding dissolved hydroxide ions:

$$NaOH \rightarrow Na^{+}_{(aq)} + OH^{-}_{(aq)}$$

Brønsted–Lowry Acid–base Theory

The Brønsted–Lowry theory is an acid–base reaction theory which was proposed independently by Johannes Nicolaus Brønsted and Thomas Martin Lowry in 1923. The fundamental concept of this theory is that when an acid and a base react with each other, the acid forms its conjugate base, and the base forms its conjugate acid by exchange of a proton (the hydrogen cation, or H^+). This theory is a generalization of the Arrhenius theory.

Definitions of Acids and Bases

Johannes Nicolaus Brønsted and Thomas Martin Lowry, independently, formulated the idea that acids are proton (H^+) donors while bases are proton acceptors.

In the Arrhenius theory acids are defined as substances which dissociate in aqueous solution to give H^+ (hydrogen ions). Bases are defined as substances which dissociate in aqueous solution to give OH^- (hydroxide ions).

In 1923 physical chemists Johannes Nicolaus Brønsted in Denmark and Thomas Martin Lowry in England independently proposed the theory that carries their names. In the Brønsted–Lowry theory acids and bases are defined by the way they react with each other, which allows for greater generality. The definition is expressed in terms of an equilibrium expression,

acid + base ⇌ conjugate base + conjugate acid.

With an acid, HA, the equation can be written symbolically as:

$$HA + B \rightleftharpoons A^- + HB^+$$

The equilibrium sign, ⇌, is used because the reaction can occur in both forward and backward directions. The acid, HA, can lose a proton to become its conjugate base, A^-. The base, B, can accept a proton to become its conjugate acid, HB^+. Most acid-base reactions are fast so that the components of the reaction are usually in dynamic equilibrium with each other.

Aqueous Solutions

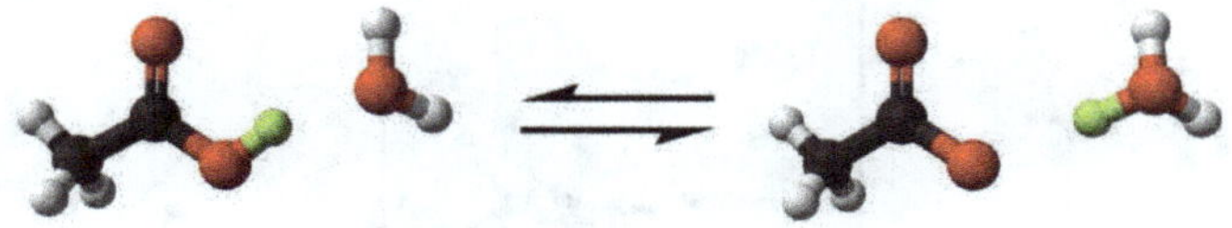

Acetic acid, a weak acid, donates a proton (hydrogen ion, highlighted in green) to water in an equilibrium reaction to give the acetate ion and the hydronium ion. Red: oxygen, black: carbon, white: hydrogen.

Consider the following acid–base reaction:

$$CH_3COOH + H_2O \rightleftharpoons CH_3COO^- + H_3O^+$$

Acetic acid, CH_3COOH, is an acid because it donates a proton to water (H_2O) and becomes its conjugate base, the acetate ion (CH_3COO^-). H_2O is a base because it accepts a proton from CH_3COOH and becomes its conjugate acid, the hydronium ion, (H_3O^+).

The reverse of an acid-base reaction is also an acid-base reaction, between the conjugate acid of the base in the first reaction and the conjugate base of the acid. In the above example, acetate is the base of the reverse reaction and hydronium ion is the acid.

$$H_3O^+ + CH_3COO^- \rightleftharpoons CH_3COOH + H_2O$$

The power of the Brønsted–Lowry theory is that, in contrast to Arrhenius theory, it does not require an acid to dissociate.

Amphoteric Substances

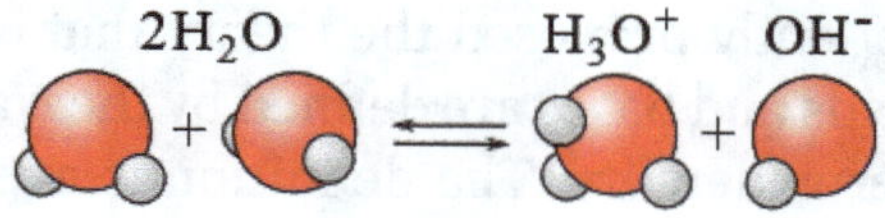

The amphoteric nature of water

The essence of Brønsted–Lowry theory is that an acid only exists as such in relation to a base, and *vice versa*. Water is amphoteric as it can act as an acid or as a base. In the image shown at the right one molecule of H_2O acts as a base and gains H^+ to become H_3O^+ while the other acts as an acid and loses H^+ to become OH^-.

Another example is furnished by substances like aluminium hydroxide, $Al(OH)_3$.

$$Al(OH)_3 + OH^- \rightleftharpoons Al(OH)_4^-\text{, acting as an acid}$$

$$3H^+ + Al(OH)_3 \rightleftharpoons 3H_2O + Al^{3+}(aq)\text{, acting as a base}$$

Non-aqueous Solutions

The hydrogen ion, or hydronium ion, is a Brønsted–Lowry acid in aqueous solutions, and the hydroxide ion is a base, by virtue of the self-dissociation reaction,

$$H_2O + H_2O \rightleftharpoons H_3O^+ + OH^-$$

An analogous reaction occurs in liquid ammonia,

$$NH_3 + NH_3 \rightleftharpoons NH_4^+ + NH_2^-$$

Thus, the ammonium ion, NH_4^+, plays the same role in liquid ammonia as does the hydronium ion in water and the amide ion, NH_2^-, is analogous to the hydroxide ion. Ammonium salts behave as acids, and amides behave as bases.

Some non-aqueous solvents can behave as bases, that is, proton acceptors, in relation to Brønsted–Lowry acids.

$$HA + S \rightleftharpoons A^- + SH^+$$

where S stands for a solvent molecule. The most important such solvents are dimethylsulfoxide, DMSO, and acetonitrile, CH_3CN, as these solvents has been widely used to measure the acid dissociation constants of organic molecules. Because DMSO is a stronger proton acceptor than H_2O the acid becomes a stronger acid in this solvent than in water. Indeed, many molecules behave as acids in non-aqueous solution that do not do so in aqueous solution. An extreme case occurs with carbon acids, where a proton is extracted from a C-H bond.

Some non-aqueous solvents can behave as acids. An acidic solvent will increase basicity of substances dissolved in it. For example, the compound CH_3COOH is known as acetic acid because of its acidic behaviour in water. However it behaves as a base in liquid hydrogen chloride, a much more acidic solvent.

$$HCl + CH_3COOH \rightleftharpoons Cl^- + CH_3C(OH)_2^+$$

Comparison with Lewis Acid–base Theory

In the same year that Brønsted and Lowry published their theory, G. N. Lewis proposed an alternative theory of acid–base reactions. The Lewis theory is based on electronic structure. A Lewis base is defined as a compound that can donate an electron pair to a Lewis acid, a compound that can accept an electron pair. Lewis's proposal gives an explanation to the Brønsted–Lowry classification in terms of electronic structure.

$$HA + B\!: \rightleftharpoons A\!:^{-} + BH^{+}$$

In this representation both the base, B, and the conjugate base, A⁻, are shown carrying a lone pair of electrons and the proton, which is a Lewis acid, is transferred between them.

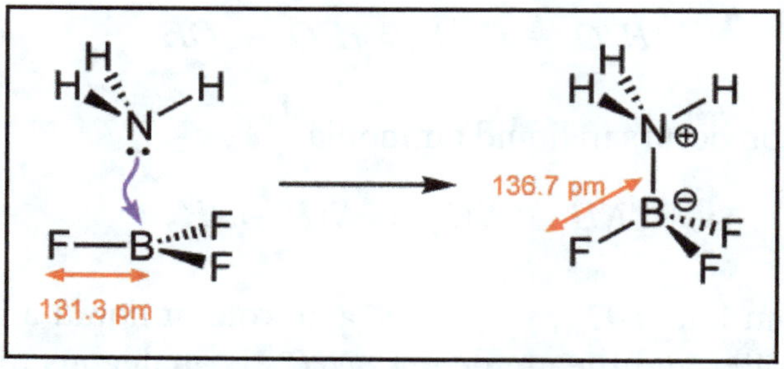

Adduct of ammonia and boron trifluoride.

Lewis later wrote in "To restrict the group of acids to those substances that contain hydrogen interferes as seriously with the systematic understanding of chemistry as would the restriction of the term oxidizing agent to substances containing oxygen." In Lewis theory an acid, A, and a base, B:, form an adduct, AB, in which the electron pair is used to form a dative covalent bond between A and B. This is illustrated with the formation of the adduct $H_3N{-}BF_3$ from ammonia and boron trifluoride, a reaction that cannot occur in aqueous solution because boron trifluoride reacts violently with water in a hydrolysis reaction.

$$BF_3 + 3H_2O \rightarrow B(OH)_3 + 3HF$$

$$HF \rightleftharpoons H^{+} + F^{-}$$

These reactions illustrate that BF_3 is an acid in both Lewis and Brønsted–Lowry classifications and emphasizes the consistency between both theories.

Boric acid is recognized as a Lewis acid by virtue of the reaction.

$$B(OH)_3 + H_2O \rightleftharpoons B(OH)_4^{-} + H^{+}$$

In this case the acid does not dissociate, it is the base, H_2O that dissociates. A solution of $B(OH)_3$ is acidic because hydrogen ions are liberated in this reaction.

There is strong evidence that dilute aqueous solutions of ammonia contain negligible amounts of the ammonium ion.

$$H_2O + NH_3 \rightarrow OH^- + NH_4^+$$

and that, when dissolved in water, ammonia functions as a Lewis base.

Comparison with the Lux-Flood Theory

The reactions between certain oxides in non-aqueous media cannot be explained on the basis of Brønsted–Lowry theory. For example, the reaction,

$$2MgO + SiO_2 \rightarrow Mg_2SiO_4$$

does not fall within the scope of the Brønsted–Lowry definition of acids and bases. On the other hand, MgO is basic and SiO_2 is acidic in the Brønsted–Lowry sense, referring to mixtures in water.

$$2H^+ + MgO(s) \rightarrow Mg^{2+}(aq) + 2H_2O$$

$$SiO_2(s) + 2H_2O \rightarrow SiO_4^{4-} + 4H^+(\equiv Si(OH)_4(aq))$$

Lux-Flood theory also classifies magnesium oxide as a base in non-aqueous circumstances. This classification is important in geochemistry. Minerals such as olivine, $(Mg,Fe)SiO_4$ are classed as ultramafic; olivine is a compound of a very basic oxide, MgO, with an acidic oxide, silica, SiO_2.

Acid Dissociation Constant

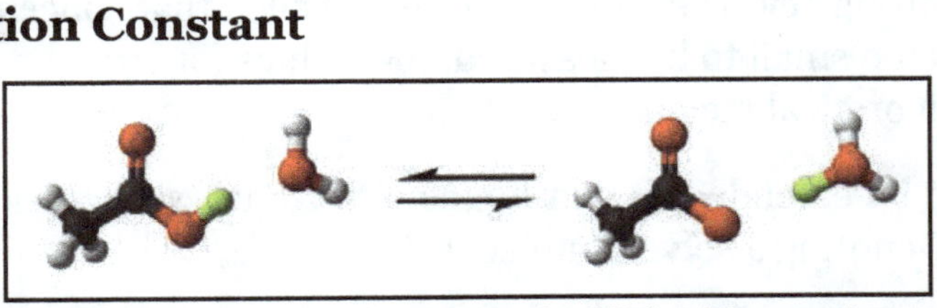

Acetic acid, a weak acid, donates a proton (hydrogen ion, highlighted in green) to water in an equilibrium reaction to give the acetate ion and the hydronium ion. Red: oxygen, black: carbon, white: hydrogen.

An acid dissociation constant, K_a, (also known as acidity constant, or acid-ionization constant) is a quantitative measure of the strength of an acid in solution. It is the equilibrium constant for a chemical reaction known as dissociation in the context of acid–base reactions. In aqueous solution, the equilibrium of acid dissociation can be written symbolically as:

$$HA + H_2O \rightleftharpoons A^- + H_3O^+$$

where HA is a generic acid that dissociates into A^-, known as the conjugate base of the acid and a hydrogen ion which combines with a water molecule to make a hydronium ion. In the example shown in the figure, HA represents acetic acid, and A^- represents the acetate ion, the conjugate base.

The chemical species HA, A^- and H_3O^+ are said to be in equilibrium when their concentrations do not change with the passing of time. The dissociation constant is usually written as a quotient of the equilibrium concentrations (in mol/L), denoted by [HA], [A⁻] and [H_3O^+],

$$K_a = \frac{[A^-][H_3O^+]}{[HA][H_2O]}$$

In all but the most concentrated aqueous solutions of an acid the concentration of water can be taken as constant and can be ignored. The definition can then be written more simply,

$$HA \rightleftharpoons A^- + H^+ : K_a = \frac{[A^-][H^+]}{[HA]}$$

This is the definition in common usage. For many practical purposes it is more convenient to discuss the logarithmic constant, pK_a,

$$pK_a = -\log_{10} K_a$$

The larger the value of pK_a, the smaller the extent of dissociation at any given pH—that is, the weaker the acid. A weak acid has a pK_a value in the approximate range −2 to 12 in water. Acids with a pK_a value of less than about −2 are said to be strong acids; the dissociation of a strong acid is effectively complete such that concentration of the undissociated acid is too small to be measured. pK_a values for strong acids can, however, be estimated by theoretical means.

The definition can be extended to non-aqueous solvents, such as acetonitrile and dimethylsulfoxide. Denoting a solvent molecule by S,

$$HA + S \rightleftharpoons A^- + SH^+ ; K_a = \frac{[A^-][SH^+]}{[HA][S]}$$

When the concentration of solvent molecules can be taken to be constant, $K_a = \frac{[A^-][H^+]}{[HA]}$, as before.

Theoretical Background

The acid dissociation constant for an acid is a direct consequence of the underlying thermodynamics of the dissociation reaction; the pK_a value is directly proportional to the standard Gibbs free energy change for the reaction. The value of the pK_a changes with temperature and can be understood qualitatively based on Le Châtelier's principle: when the reaction is endothermic, K_a increases and pK_a decreases with increasing temperature; the opposite is true for exothermic reactions.

The value of pK_a also depends on molecular structure of the acid in many ways. For example, Pauling proposed two rules: one for successive pK_a of polyprotic acids, and one to estimate the pK_a of oxyacids based on the number of =O and −OH groups. Other structural factors that influence the magnitude of the acid dissociation constant include inductive effects, mesomeric effects, and hydrogen bonding. Hammett type equations have frequently been applied to the estimation of pK_a.

The quantitative behaviour of acids and bases in solution can be understood only if their pK_a values are known. In particular, the pH of a solution can be predicted when the analytical concentration and pK_a values of all acids and bases are known; conversely, it is possible to calculate the equilibrium concentration of the acids and bases in solution when the pH is known. These calculations find application in many different areas of chemistry, biology, medicine, and geology. For example, many compounds used for medication are weak acids or bases, and a knowledge of the pK_a values, together with the water–octanol partition coefficient, can be used for estimating the extent to which the compound enters the blood stream. Acid dissociation constants are also essential in aquatic chemistry and chemical oceanography, where the acidity of water plays a fundamental role. In living organisms, acid–base homeostasis and enzyme kinetics are dependent on the pK_a values of the many acids and bases present in the cell and in the body. In chemistry, a knowledge of pK_a values is necessary for the preparation of buffer solutions and is also a prerequisite for a quantitative understanding of the interaction between acids or bases and metal ions to form complexes. Experimentally, pK_a values can be determined by potentiometric (pH) titration, but for values of pK_a less than about 2 or more than about 11, spectrophotometric or NMR measurements may be required due to practical difficulties with pH measurements.

Definitions

According to Arrhenius's original definition, an acid is a substance that dissociates in aqueous solution, releasing the hydrogen ion H^+ (a proton):

$$HA \rightleftharpoons A^- + H^+ .$$

The equilibrium constant for this dissociation reaction is known as a dissociation constant. The liberated proton combines with a water molecule to give a hydronium (or oxonium) ion H_3O^+ (naked protons do not exist in solution), and so Arrhenius later proposed that the dissociation should be written as an acid–base reaction:

$$HA + H_2O \rightleftharpoons A^- + H_3O^+$$

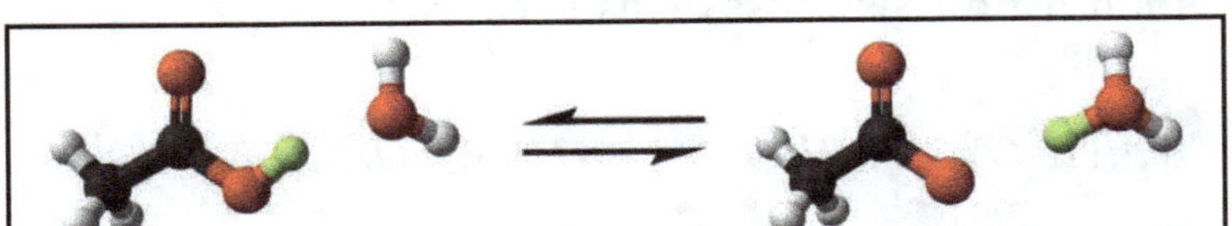

Acetic acid, a weak acid, donates a proton (hydrogen ion, highlighted in green) to water in an equilibrium reaction to give the acetate ion and the hydronium ion. Red: oxygen, black: carbon, white: hydrogen.

Brønsted and Lowry generalised this further to a proton exchange reaction:

acid + base ⇌ conjugate base + conjugate acid.

The acid loses a proton, leaving a conjugate base; the proton is transferred to the base, creating a conjugate acid. For aqueous solutions of an acid HA, the base is water; the conjugate base is A^- and the conjugate acid is the hydronium ion. The Brønsted–Lowry definition applies to other solvents, such as dimethyl sulfoxide: the solvent S acts as a base, accepting a proton and forming the conjugate acid SH^+.

$$HA + S \rightleftharpoons A^- + SH^+.$$

In solution chemistry, it is common to use H^+ as an abbreviation for the solvated hydrogen ion, regardless of the solvent. In aqueous solution H^+ denotes a solvated hydronium ion rather than a proton.

The designation of an acid or base as "conjugate" depends on the context. The conjugate acid BH^+ of a base B dissociates according to,

$$BH^+ + OH^- \rightleftharpoons B + H_2O$$

which is the reverse of the equilibrium

H_2O (acid) + B (base) ⇌ OH^- (conjugate base) + BH^+ (conjugate acid).

The hydroxide ion OH^-, a well known base, is here acting as the conjugate base of the acid water. Acids and bases are thus regarded simply as donors and acceptors of protons respectively.

A broader definition of acid dissociation includes hydrolysis, in which protons are produced by the splitting of water molecules. For example, boric acid ($B(OH)_3$) produces H_3O^+ as if it were a proton donor, but it has been confirmed by Raman spectroscopy that this is due to the hydrolysis equilibrium:

$$B(OH)_3 + 2\,H_2O \rightleftharpoons B(OH)_4^- + H_3O^+.$$

Similarly, metal ion hydrolysis causes ions such as $[Al(H_2O)_6]^{3+}$ to behave as weak acids:

$$[Al(H_2O)_6]^{3+} + H_2O \rightleftharpoons [Al(H_2O)_5(OH)]^{2+} + H_3O^+.$$

According to Lewis's original definition, an acid is a substance that accepts an electron pair to form a coordinate covalent bond.

Equilibrium Constant

An acid dissociation constant is a particular example of an equilibrium constant. For the specific equilibrium between a monoprotic acid, HA and its conjugate base A^-, in water,

$$HA + H_2O \rightleftharpoons A^- + H_3O^+$$

the thermodynamic equilibrium constant, $K^\ominus$ can be defined by,

$$= \frac{\{A^-\}\{H_3O^+\}}{\{HA\}\{H_2O\}}$$

where {A} is the activity of the chemical species A etc. $K^\ominus$ is dimensionless since activity is dimensionless. Activities of the products of dissociation are placed in the numerator, activities of the reactants are placed in the denominator. See activity coefficient for a derivation of this expression.

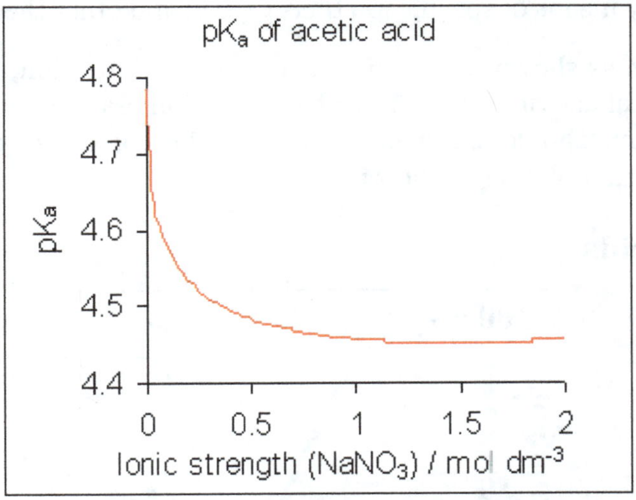

Variation of pK_a of acetic acid with ionic strength.

Since activity is the product of concentration and activity coefficient (γ) the definition could also be written as,

$$K^\ominus = \frac{[A^-][H_3O^+]}{[HA][H_2O]} \times \frac{\gamma_{A^-}\,\gamma_{H_3O^+}}{\gamma_{HA}\,\gamma_{H_2O}} = \frac{[A^-][H_3O^+]}{[HA][H_2O]} \times \Gamma$$

where [HA] represents the concentration of HA and Γ is a quotient of activity coefficients.

To avoid the complications involved in using activities, dissociation constants are determined, where possible, in a medium of high ionic strength, that is, under conditions in which Γ can be assumed to be always constant. For example, the medium might be a solution of 0.1 M sodium nitrate or 3 M potassium perchlorate (1 M = 1 mol·dm^{-3}, a unit of molar concentration). Furthermore, in all but the most concentrated solutions it can be assumed that the concentration of water, [H_2O], is constant, approximately 55 mol·dm^{-3}. On dividing $K^{\ominus}$ by the constant terms and writing [H^+] for the concentration of the hydronium ion the expression,

$$K_a = \frac{[A^-][H^+]}{[HA]}$$

is obtained. This is the definition in common use. pK_a is defined as $-\log_{10} K_a$. Note, however, that all published dissociation constant values refer to the specific ionic medium used in their determination and that different values are obtained with different conditions, as shown for acetic acid in the illustration above. When published constants refer to an ionic strength other than the one required for a particular application, they may be adjusted by means of specific ion theory (SIT) and other theories.

Using the equation as shown K_a has dimensions of concentration, but the exact definition uses chemical activities, which can be dimensionless. Therefore, K_a, as defined properly, is also dimensionless. But as defined here it is correct to quote a value with a dimension as, for example, "K_a = 300 M".

Monoprotic Acids

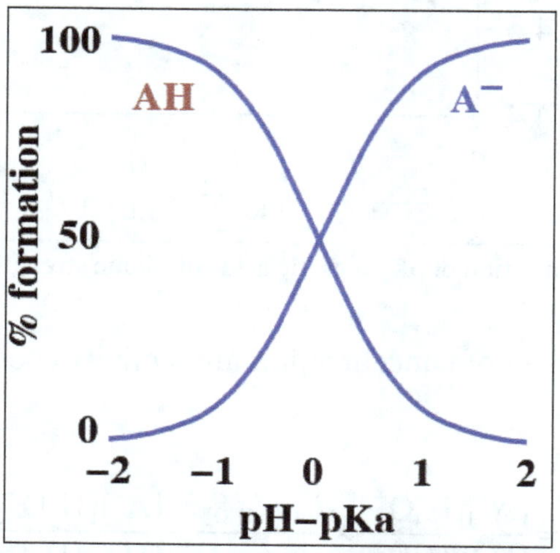

Variation of the % formation of a monoprotic acid, AH, and its conjugate base, A$^-$, with the difference between the pH and the pK_a of the acid.

After rearranging the expression defining K_a, and putting pH = $-\log_{10}[H^+]$, one obtains

$$pH = pK_a + \log\frac{[A^-]}{[HA]}$$

This is a form of the Henderson–Hasselbalch equation, from which the following conclusions can be drawn.

$$pH - pK_a = \log\frac{[A^-]}{[HA]}$$

- At half-neutralization $[A^-]/[HA] = 1$; since $\log(1) = 0$, the pH at half-neutralization is numerically equal to pK_a. Conversely, when pH = pK_a, the concentration of HA is equal to the concentration of A^-.
- The buffer region extends over the approximate range $pK_a \pm 2$, though buffering is weak outside the range $pK_a \pm 1$. At $pK_a \pm 1$, $[A^-]/[HA] = 10$ or $1/10$.
- If the pH is known, the ratio may be calculated. This ratio is independent of the analytical concentration of the acid.

In water, measurable pK_a values range from about −2 for a strong acid to about 12 for a very weak acid (or strong base). All acids with a pK_a value of less than −2 are more than 99% dissociated at pH 0 (1 M acid). This is known as solvent leveling since all such acids are brought to the same level of being *strong acids*, regardless of their pK_a values. Likewise, all bases with a pK_a value larger than the upper limit are more than 99% *protonated* at all attainable pH values and are classified as *strong bases*.

An example of a strong acid is hydrochloric acid, HCl, which has a pK_a value, estimated from thermodynamic quantities, of −9.3 in water. The concentration of undissociated acid in a 1 mol·dm^{-3} solution will be less than 0.01% of the concentrations of the products of dissociation. Hydrochloric acid is said to be "fully dissociated" in aqueous solution because the amount of undissociated acid is imperceptible. When the pK_a and analytical concentration of the acid are known, the extent of dissociation and pH of a solution of a monoprotic acid can be easily calculated using an ICE table.

A buffer solution of a desired pH can be prepared as a mixture of a weak acid and its conjugate base. In practice the mixture can be created by dissolving the acid in water, and adding the requisite amount of strong acid or base. The pK_a of the acid must be less than two units different from the target pH.

Polyprotic Acids

Polyprotic acids are acids that can lose more than one proton. The constant for dissociation of the first proton may be denoted as K_{a1} and the constants for dissociation of successive protons as K_{a2}, etc. Phosphoric acid, H_3PO_4, is an example of a polyprotic acid as it can lose three protons.

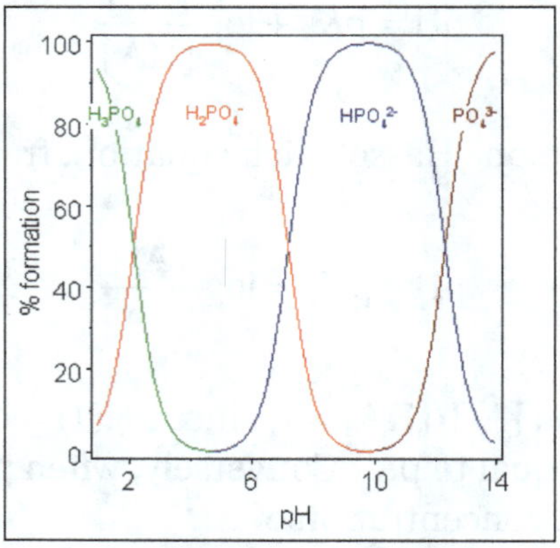

Phosphoric acid speciation.

Equilibrium	pK_a value
$H_3PO_4 \rightleftharpoons H_2PO_4^- + H^+$	$pK_{a1} = 2.14$
$H_2PO_4^- \rightleftharpoons HPO_4^{2-} + H^+$	$pK_{a2} = 7.20$
$HPO_4^{2-} \rightleftharpoons PO_4^{3-} + H^+$	pK_{a3} 12.37

When the difference between successive pK values is about four or more, as in this example, each species may be considered as an acid in its own right; In fact salts of $H_2PO_4^-$ may be crystallised from solution by adjustment of pH to about 5.5 and salts of HPO_4^{2-} may be crystallised from solution by adjustment of pH to about 10. The species distribution diagram shows that the concentrations of the two ions are maximum at pH 5.5 and 10.

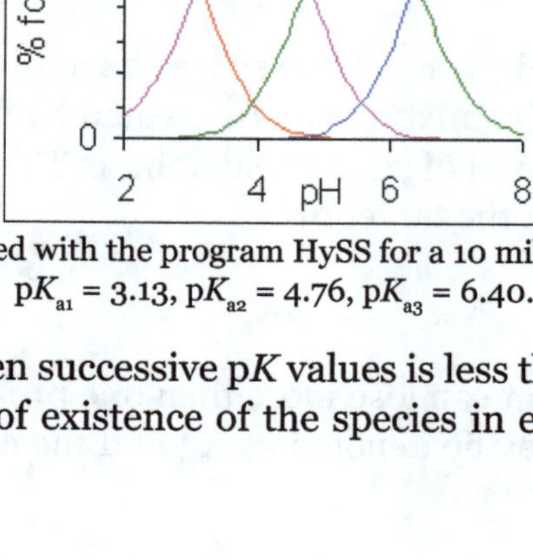

% species formation calculated with the program HySS for a 10 millimolar solution of citric acid. $pK_{a1} = 3.13$, $pK_{a2} = 4.76$, $pK_{a3} = 6.40$.

When the difference between successive pK values is less than about four there is overlap between the pH range of existence of the species in equilibrium. The smaller the difference, the more the overlap. The case of citric acid is shown; solutions of citric acid are buffered over the whole range of pH 2.5 to 7.5.

According to Pauling's first rule, successive p*K* values of a given acid increase ($pK_{a2} > pK_{a1}$). For oxyacids with more than one ionizable hydrogen on the same atom, the pK_a values often increase by about 5 units for each proton removed, as in the example of phosphoric acid above.

In the case of a diprotic acid, H_2A, the two equilibria are

$$H_2A \rightleftharpoons HA^- + H^+$$

$$HA^- \rightleftharpoons A^{2-} + H^+$$

it can be seen that the second proton is removed from a negatively charged species. Since the proton carries a positive charge extra work is needed to remove it; that is the cause of the trend noted above. Phosphoric acid values (above) illustrate this rule, as do the values for vanadic acid, H_3VO_4. When an exception to the rule is found it indicates that a major change in structure is occurring. In the case of VO_2^+ (aq), the vanadium is octahedral, 6-coordinate, whereas vanadic acid is tetrahedral, 4-coordinate. This is the basis for an explanation of why $pK_{a1} > pK_{a2}$ for vanadium(V) oxoacids.

Equilibrium	pK_a value
$[VO_2(H_2O)_4]^+ \rightleftharpoons H_3VO_4 + H^+ + 2H_2O$	$pK_{a1} = 4.2$
$H_3VO_4 \rightleftharpoons H_2VO_4^- + H^+$	$pK_{a2} = 2.60$
$H_2VO_4^- \rightleftharpoons HVO_4^{2-} + H^+$	$pK_{a3} = 7.92$
$HVO_4^{2-} \rightleftharpoons VO_4^{3-} + H^+$	$pK_{a4} = 13.27$

Isoelectric Point

For substances in solution the isoelectric point (p*I*) is defined as the pH at which the sum, weighted by charge value, of concentrations of positively charged species is equal to the weighted sum of concentrations of negatively charged species. In the case that there is one species of each type, the isoelectric point can be obtained directly from the p*K* values. Take the example of glycine, defined as AH. There are two dissociation equilibria to consider.

$$AH_2^+ \rightleftharpoons AH + H^+; \quad [AH][H^+] = K_1[AH_2^+]$$

$$AH \rightleftharpoons A^- + H^+; \quad [A^-][H^+] = K_2[AH]$$

Substitute the expression for [AH] into the first equation,

$$[A^-][H^+]^2 = K_1K_2[AH_2^+]$$

At the isoelectric point the concentration of the positively charged species, AH_2^+, is equal to the concentration of the negatively charged species, A^-, so

$$[H^+]^2 = K_1K_2$$

Therefore, taking cologarithms, the pH is given by,

$$pI = \frac{pK_1 + pK_2}{2}$$

pI values for amino acids are listed at Proteinogenic amino acid#Chemical properties. When more than two charged species are in equilibrium with each other a full speciation calculation may be needed.

Water Self-ionization

Water possesses both acidic and basic properties and is said to be amphiprotic. The ionization equilibrium can be written,

$$H_2O \rightleftharpoons OH^- + H^+$$

where in aqueous solution H^+ or H^+(aq) denotes a solvated proton. Often this is written as the hydronium ion H_3O^+, but this formula is not exact because in fact there is solvation by more than one water molecule and species such as $H_5O_2^+$, $H_7O_3^+$ and $H_9O_4^+$ are also present.

The equilibrium constant is given by,

$$K_a = \frac{[H^+][OH^-]}{[H_2O]}$$

When, as is usually the case, the concentration of water can be assumed to be constant, this expression may be replaced by,

$$K_w = [H^+][OH^-]$$

The self-ionization constant of water, K_w, is thus just a special case of an acid dissociation constant. A logarithmic form analogous to pK_a may also be defined,

$$pK_w = -\log_{10}(K_w)$$

pK_w values for pure water at various temperatures											
***T*/°C**	0	5	10	15	20	25	30	35	40	45	50
pK_w	14.943	14.734	14.535	14.346	14.167	13.997	13.830	13.680	13.535	13.396	13.262

These data can be fitted to a parabola with,

$$pK_w = 14.94 - 0.04209T + 0.0001718T^2$$

From this equation, $pK_w = 14$ at 24.87 °C. At that temperature both hydrogen and hydroxide ions have a concentration of 10^{-7} mol dm^{-3}.

Protonation Constants

The dissociation of a monoprotic acid can also be described as the protonation of the conjugate base of the acid,

$$A^- + H^+ \rightleftharpoons AH$$

This leads to the definition of a *association* (protonation) constant, denoted here as $K_{association}$, as,

$$K_{association} = \frac{[HA]}{[A^-][H^+]}$$

The *dissociation* (deprotonation) constant definition can be written as,

$$K_{dissociation} = \frac{[A^-][H^+]}{[HA]}$$

The definitions show that the values of the two constants are reciprocals of each other, and

$$pK_{dissociation} = \log K_{association}$$

The situation is a little more complicated with polybasic acids. For example, with phosphoric acid,

$$pK_{a1} = \log K(\text{association})_3$$

$$pK_{a2} = \log K(\text{association})_2$$

$$pK_{a3} = \log K(\text{association})_1$$

Amphoteric Substances

An amphoteric substance is one that can act as an acid or as a base, depending on pH. Water (above) is amphoteric. Another example of an amphoteric molecule is the bicarbonate ion HCO_3^- that is the conjugate base of the carbonic acid molecule H_2CO_3 in the equilibrium,

$$H_2CO_3 + H_2O \rightleftharpoons HCO_3^- + H_3O^+$$

but also the conjugate acid of the carbonate ion CO_3^{2-} in (the reverse of) the equilibrium,

$$HCO_3^- + OH^- \rightleftharpoons CO_3^{2-} + H_2O.$$

Carbonic acid equilibria are important for acid–base homeostasis in the human body.

An amino acid is also amphoteric with the added complication that the neutral molecule is subject to an internal acid–base equilibrium in which the basic amino group attracts and binds the proton from the acidic carboxyl group, forming a zwitterion.

$$NH_2CHRCO_2H \rightleftharpoons NH_3^+CHRCO_2^-$$

At pH less than about 5 both the carboxylate group and the amino group are protonated. As pH increases the acid dissociates according to,

$$NH_3^+CHRCO_2H \rightleftharpoons NH_3^+CHRCO_2^- + H^+$$

At high pH a second dissociation may take place.

$$NH_3^+CHRCO_2^- \rightleftharpoons NH_2CHRCO_2^- + H^+$$

Thus the zwitterion, $NH_3^+CHRCO_2^-$, is amphoteric because it may either be protonated or deprotonated.

Bases and Basicity

The equilibrium constant K_b for a base is usually defined as the *association* constant for protonation of the base, B, to form the conjugate acid, HB^+.

$$B + H_2O \rightleftharpoons HB^+ + OH^-$$

Using similar reasoning to that used before,

$$K_b = \frac{[HB^+][OH^-]}{[B]}$$

$$pK_b = -\log_{10}(K_b)$$

K_b is related to K_a for the conjugate acid. In water, the concentration of the hydroxide ion, $[OH^-]$, is related to the concentration of the hydrogen ion by $K_w = [H^+][OH^-]$, therefore,

$$[OH^-] = \frac{K_w}{[H^+]}$$

Substitution of the expression for $[OH^-]$ into the expression for K_b gives,

$$K_b = \frac{[HB^+]K_w}{[B][H^+]} = \frac{K_w}{K_a}$$

When K_a, K_b and K_w are determined under the same conditions of temperature and ionic strength, it follows, taking cologarithms, that pK_b = pK_w – pK_a. In aqueous solutions at 25 °C, pK_w is 13.9965, so,

$$pK_b \approx 14 - pK_a$$

with sufficient accuracy for most practical purposes. In effect there is no need to define pK_b separately from pK_a, but it is done here as often only pK_b values can be found in the older literature.

For metal hydroxides K_b can also be defined as the *dissociation* constant for loss of a hydroxide ion: $B(OH) \rightleftharpoons B^+ + OH^-$ or $B(OH)_2 \rightleftharpoons B(OH)^+ + OH^-$. This is the reciprocal of a stability constant for formation of the complex.

Temperature Dependence

All equilibrium constants vary with temperature according to the Van't Hoff equation,

$$\frac{d\ln(K)}{dT} = \frac{\Delta H^\ominus}{RT^2}$$

R is the gas constant and T is the absolute temperature. Thus, for exothermic reactions, (the standard enthalpy change, ΔH, is negative) K decreases with temperature, but for endothermic reactions (ΔH is positive) K increases with temperature.

The standard enthalpy change for a reaction is itself a function of temperature, according to Kirchhoff's law of thermochemistry:

$$\left(\frac{\partial \Delta H}{\partial T}\right)_p = \Delta C_p$$

where ΔC_p is the heat capacity change at constant pressure. In practice ΔH may be taken to be constant over a small temperature range.

Acidity in Nonaqueous Solutions

A solvent will be more likely to promote ionization of a dissolved acidic molecule in the following circumstances:

1. It is a protic solvent, capable of forming hydrogen bonds.
2. It has a high donor number, making it a strong Lewis base.

3. It has a high dielectric constant (relative permittivity), making it a good solvent for ionic species.

pK_a values of organic compounds are often obtained using the aprotic solvents dimethyl sulfoxide (DMSO) and acetonitrile (ACN).

Solvent properties at 25 °C		
Solvent	**Donor number**	**Dielectric constant**
Acetonitrile	14	37
Dimethylsulfoxide	30	47
Water	18	78

DMSO is widely used as an alternative to water because it has a lower dielectric constant than water, and is less polar and so dissolves non-polar, hydrophobic substances more easily. It has a measurable pK_a range of about 1 to 30. Acetonitrile is less basic than DMSO, and, so, in general, acids are weaker and bases are stronger in this solvent. Some pK_a values at 25 °C for acetonitrile (ACN) and dimethyl sulfoxide (DMSO) are shown in the following tables. Values for water are included for comparison.

pK_a values of acids			
$HA \rightleftharpoons A^- + H^+$	**ACN**	**DMSO**	**water**
p-Toluenesulfonic acid	8.5	0.9	strong
2,4-Dinitrophenol	16.66	5.1	3.9
Benzoic acid	21.51	11.1	4.2
Acetic acid	23.51	12.6	4.756
Phenol	29.14	18.0	9.99
$BH^+ \rightleftharpoons B + H^+$	**ACN**	**DMSO**	**water**
Pyrrolidine	19.56	10.8	11.4
Triethylamine	18.82	9.0	10.72
Proton sponge	18.62	7.5	12.1
Pyridine	12.53	3.4	5.2
Aniline	10.62	3.6	4.6

Ionization of acids is less in an acidic solvent than in water. For example, hydrogen chloride is a weak acid when dissolved in acetic acid. This is because acetic acid is a much weaker base than water.

$$HCl + CH_3CO_2H \rightleftharpoons Cl^- + CH_3C(OH)_2^+$$

$$\text{acid} + \text{base} \rightleftharpoons \text{conjugate base} + \text{conjugate acid}$$

Compare this reaction with what happens when acetic acid is dissolved in the more acidic solvent pure sulfuric acid,

$$H_2SO_4 + CH_3CO_2H \rightleftharpoons HSO_4^- + CH3C(OH) + 2$$

The unlikely geminal diol species $CH_3C(OH)_2^+$ is stable in these environments. For aqueous solutions the pH scale is the most convenient acidity function. Other acidity functions have been proposed for non-aqueous media, the most notable being the Hammett acidity function, H_o, for superacid media and its modified version H_- for superbasic media.

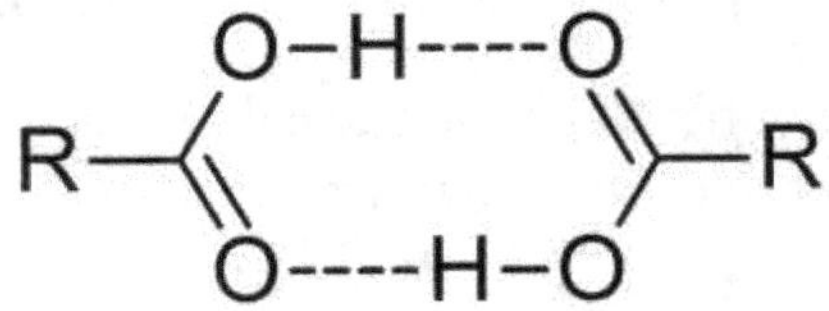

Dimerization of a carboxylic acid.

In aprotic solvents, oligomers, such as the well-known acetic acid dimer, may be formed by hydrogen bonding. An acid may also form hydrogen bonds to its conjugate base. This process, known as homoconjugation, has the effect of enhancing the acidity of acids, lowering their effective pK_a values, by stabilizing the conjugate base. Homoconjugation enhances the proton-donating power of toluenesulfonic acid in acetonitrile solution by a factor of nearly 800. In aqueous solutions, homoconjugation does not occur, because water forms stronger hydrogen bonds to the conjugate base than does the acid.

Mixed Solvents

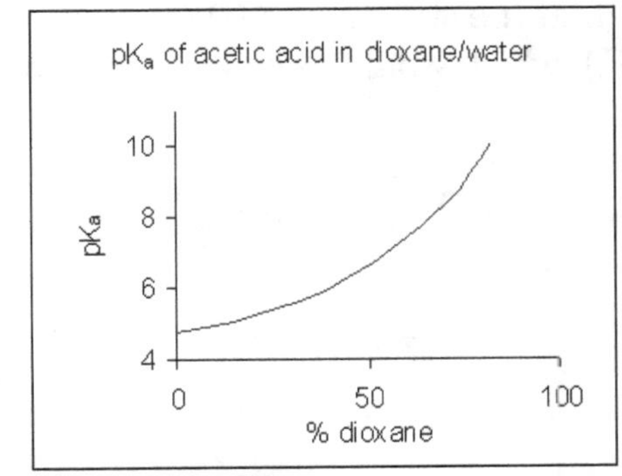

pK_a of acetic acid in dioxane/water mixtures. Data at 25 °C from Pine *et al.*

When a compound has limited solubility in water it is common practice (in the pharmaceutical industry, for example) to determine pK_a values in a solvent mixture such as water/dioxane or water/methanol, in which the compound is more soluble. In the example shown at the right, the pK_a value rises steeply with increasing percentage of dioxane as the dielectric constant of the mixture is decreasing.

A pK_a value obtained in a mixed solvent cannot be used directly for aqueous solutions. The reason for this is that when the solvent is in its standard state its activity is *defined* as one. For example, the standard state of water:dioxane 9:1 is precisely that solvent mixture, with no added solutes. To obtain the pK_a value for use with aqueous solutions it has to be extrapolated to zero co-solvent concentration from values obtained from various co-solvent mixtures.

These facts are obscured by the omission of the solvent from the expression that is normally used to define pK_a, but pK_a values obtained in a *given* mixed solvent can be compared to each other, giving relative acid strengths. The same is true of pK_a values obtained in a particular non-aqueous solvent such a DMSO.

As of 2008, a universal, solvent-independent, scale for acid dissociation constants has not been developed, since there is no known way to compare the standard states of two different solvents.

Factors that Affect pK_a Values

Pauling's second rule is that the value of the first pK_a for acids of the formula $XO_m(OH)_n$ depends primarily on the number of oxo groups *m*, and is approximately independent of the number of hydroxy groups *n*, and also of the central atom X. Approximate values of pK_a are 8 for $m = 0$, 2 for $m = 1$, −3 for $m = 2$ and < −10 for $m = 3$. Alternatively, various numerical formulas have been proposed including pK_a = 8 − 5*m* (known as Bell's rule), pK_a = 7 − 5*m*, or pK_a = 9 − 7*m*. The dependence on *m* correlates with the oxidation state of the central atom, X: the higher the oxidation state the stronger the oxyacid. For example, pK_a for HClO is 7.2, for $HClO_2$ is 2.0, for $HClO_3$ is −1 and $HClO_4$ is a strong acid (p$K_a \ll 0$). This rule can help assign molecular structure: for example phosphorous acid (H_3PO_3) has a pK_a near 2 suggested that the structure is $HPO(OH)_2$, as later confirmed by NMR spectroscopy, and not $P(OH)_3$ which would be expected to have a pK_a near 8.

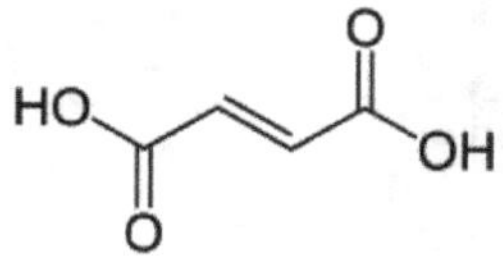

Fumaric acid.

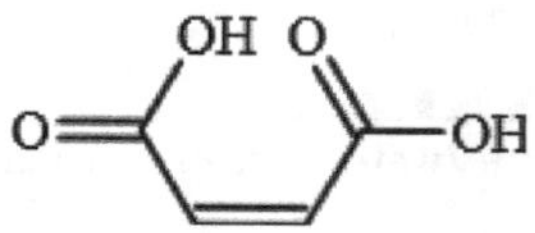

Maleic acid.

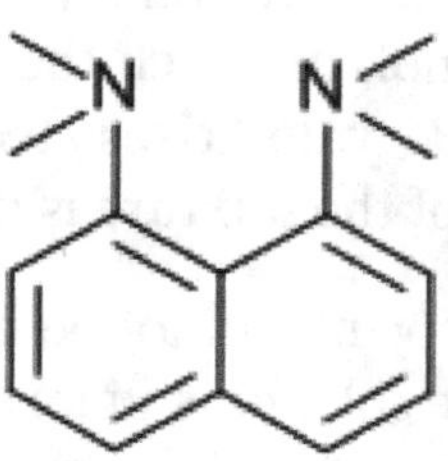

proton sponge.

With organic acids inductive effects and mesomeric effects affect the pK_a values. A simple example is provided by the effect of replacing the hydrogen atoms in acetic acid by the more electronegative chlorine atom. The electron-withdrawing effect of the substituent makes ionisation easier, so successive pK_a values decrease in the series 4.7, 2.8, 1.4 and 0.7 when 0, 1, 2 or 3 chlorine atoms are present. The Hammett equation, provides a general expression for the effect of substituents.

$$\log\left(K_a\right)=\log\left(K_a^0\right)+\rho\sigma.$$

K_a is the dissociation constant of a substituted compound, K_a^0 is the dissociation constant when the substituent is hydrogen, ρ is a property of the unsubstituted compound and σ has a particular value for each substituent. A plot of log K_a against σ is a straight line with intercept log K_a^0 and slope ρ. This is an example of a linear free energy relationship as log (K_a) is proportional to the standard fee energy change. Hammett originally formulated the relationship with data from benzoic acid with different substiuents in the *ortho-* and *para-* positions: some numerical values are in Hammett equation. This and other studies allowed substituents to be ordered according to their electron-withdrawing or electron-releasing power, and to distinguish between inductive and mesomeric effects.

Alcohols do not normally behave as acids in water, but the presence of a double bond adjacent to the OH group can substantially decrease the pK_a by the mechanism of keto–enol tautomerism. Ascorbic acid is an example of this effect. The diketone 2,4-pentanedione (acetylacetone) is also a weak acid because of the keto–enol equilibrium. In aromatic compounds, such as phenol, which have an OH substituent, conjugation with the aromatic ring as a whole greatly increases the stability of the deprotonated form.

Structural effects can also be important. The difference between fumaric acid and maleic acid is a classic example. Fumaric acid is (E)-1,4-but-2-enedioic acid, a *trans* isomer, whereas maleic acid is the corresponding *cis* isomer, i.e. (Z)-1,4-but-2-enedioic acid. Fumaric acid has pK_a values of approximately 3.0 and 4.5. By contrast, maleic acid has pK_a values of approximately 1.5 and 6.5. The reason for this large difference is that when one proton is removed from the *cis* isomer (maleic acid) a strong intramolecular hydrogen bond is formed with the nearby remaining carboxyl group. This favors the formation of the maleate H^+, and it opposes the removal of the second proton from that species. In the *trans* isomer, the two carboxyl groups are always far apart, so hydrogen bonding is not observed.

Proton sponge, 1,8-bis(dimethylamino)naphthalene, has a pK_a value of 12.1. It is one of the strongest amine bases known. The high basicity is attributed to the relief of strain upon protonation and strong internal hydrogen bonding.

Effects of the solvent and solvation should be mentioned also in this section. It turns out, these influences are more subtle than that of a dielectric medium mentioned above.

For example, the expected (by electronic effects of methyl substituents) and observed in gas phase order of basicity of methylamines, $Me_3N > Me_2NH > MeNH_2 > NH_3$, is changed by water to $Me_2NH > MeNH_2 > Me_3N > NH_3$. Neutral methylamine molecules are hydrogen-bonded to water molecules mainly through one acceptor, N–HOH, interaction and only occasionally just one more donor bond, $NH–OH_2$. Hence, methylamines are stabilized to about the same extent by hydration, regardless of the number of methyl groups. In stark contrast, corresponding methylammonium cations always utilize all the available protons for donor $NH–OH_2$ bonding. Relative stabilization of methylammonium ions thus decreases with the number of methyl groups explaining the order of water basicity of methylamines.

Thermodynamics

An equilibrium constant is related to the standard Gibbs energy change for the reaction, so for an acid dissociation constant,

$$\Delta G^{\ominus} = -RT\ln K_a \approx 2.303RT\ pK_a.$$

R is the gas constant and T is the absolute temperature. Note that $pK_a = -\log K_a$ and $2.303 \approx \ln 10$. At 25 °C $\Delta G^{\ominus}$ in kJ·mol⁻¹ = 5.708 pK_a (1 kJ·mol⁻¹ = 1000 joules per mole). Free energy is made up of an enthalpy term and an entropy term.

$$\Delta G^{\ominus} = \Delta H^{\ominus} - T\Delta S^{\ominus}$$

The standard enthalpy change can be determined by calorimetry or by using the van't Hoff equation, though the calorimetric method is preferable. When both the standard enthalpy change and acid dissociation constant have been determined, the standard entropy change is easily calculated from the equation above. In the following table, the entropy terms are calculated from the experimental values of pK_a and ΔH. The data were critically selected and refer to 25 °C and zero ionic strength, in water.

Acids					
Compound	**Equilibrium**	**pK_a**	$\Delta G^{\ominus}$ (kJ·mol⁻¹)	$\Delta H^{\ominus}$ (kJ·mol⁻¹)	$-T\Delta S^{\ominus}$ (kJ·mol⁻¹)
HA = Acetic acid	$HA \rightleftharpoons H^+ + A^-$	4.756	27.147[†]	−0.41	27.56[‡]
H_2A^+ = GlycineH⁺	$H_2A^+ \rightleftharpoons HA + H^+$	2.351	13.420	4.00	9.419
	$HA \rightleftharpoons H^+ + A^-$	9.78	55.825	44.20	11.6
H_2A = Maleic acid	$H_2A \rightleftharpoons HA^- + H^+$	1.92	10.76	1.10	9.85
	$HA^- \rightleftharpoons H^+ + A^{2-}$	6.27	35.79	−3.60	39.4
H_3A = Citric acid	$H_3A \rightleftharpoons H_2A^- + H^+$	3.128	17.855	4.07	13.78
	$H_2A^- \rightleftharpoons HA^{2-} + H^+$	4.76	27.176	2.23	24.9
	$HA^{2-} \rightleftharpoons A^{3-} + H^+$	6.40	36.509	−3.38	39.9

H_3A = Boric acid	$H_3A \rightleftharpoons H_2A^- + H^+$	9.237	52.725	13.80	38.92
H_3A = Phosphoric acid	$H_3A \rightleftharpoons H_2A^- + H^+$	2.148	12.261	−8.00	20.26
	$H_2A^- \rightleftharpoons HA^{2-} + H^+$	7.20	41.087	3.60	37.5
	$HA^{2-} \rightleftharpoons A^{3-} + H^+$	12.35	80.49	16.00	54.49
HA^- = Hydrogen sulfate	$HA^- \rightleftharpoons A^{2-} + H^+$	1.99	11.36	−22.40	33.74
H_2A = Oxalic acid	$H_2A \rightleftharpoons HA^- + H^+$	1.27	7.27	−3.90	11.15
	$HA^- \rightleftharpoons A^{2-} + H^+$	4.266	24.351	7.00	31.35

† $\Delta G^{\ominus} \approx 2.303RTpK_a$

‡ Computed here, from ΔH and ΔG values supplied in the citation, using $-T\Delta S^{\ominus} = \Delta G^{\ominus} - \Delta H^{\ominus}$

Conjugate acids of bases				
Compound	**Equilibrium**	**pK_a**	$\Delta H^{\ominus}$ (kJ·mol^{-1})	$-T\Delta S^{\ominus}$ (kJ·mol^{-1})
B = Ammonia	$HB^+ \rightleftharpoons B + H^+$	9.245	51.95	0.8205
B = Methylamine	$HB^+ \rightleftharpoons B + H^+$	10.645	55.34	5.422
B = Triethylamine	$HB^+ \rightleftharpoons B + H^+$	10.72	43.13	18.06

The first point to note is that, when pK_a is positive, the standard free energy change for the dissociation reaction is also positive. Second, some reactions are exothermic and some are endothermic, but, when $\Delta H^{\ominus}$ is negative $T\Delta S^{\ominus}$ is the dominant factor, which determines that $\Delta G^{\ominus}$ is positive. Last, the entropy contribution is always unfavourable ($(\Delta S^{\ominus} < 0)$) in these reactions. Ions in aqueous solution tend to orient the surrounding water molecules, which orders the solution and decreases the entropy. The contribution of an ion to the entropy is the partial molar entropy which is often negative, especially for small or highly charged ions. The ionization of a neutral acid involves formation of two ions so that the entropy decreases ($(\Delta S^{\ominus} < 0)$). On the second ionization of the same acid, there are now three ions and the anion has a charge, so the entropy again decreases.

Note that the *standard* free energy change for the reaction is for the changes *from* the reactants in their standard states *to* the products in their standard states. The free energy change *at* equilibrium is zero since the chemical potentials of reactants and products are equal at equilibrium.

Experimental Determination

The experimental determination of pK_a values is commonly performed by means of titrations, in a medium of high ionic strength and at constant temperature. A typical procedure would be as follows. A solution of the compound in the medium is acidified with a strong acid to the point where the compound is fully protonated. The solution is then titrated with a strong base until all the protons have been removed. At each point in the titration pH is measured using a glass electrode and a pH meter. The equilibrium constants are found by fitting calculated pH values to the observed values, using the method of least squares.

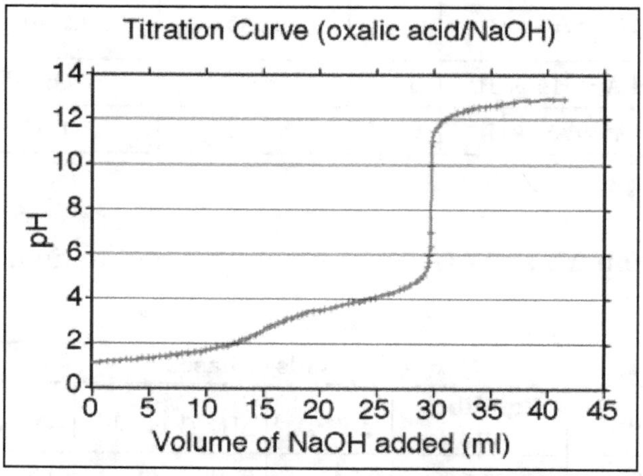

A calculated titration curve of oxalic acid titrated with a solution of sodium hydroxide.

The total volume of added strong base should be small compared to the initial volume of titrand solution in order to keep the ionic strength nearly constant. This will ensure that pK_a remains invariant during the titration.

A calculated titration curve for oxalic acid is shown. Oxalic acid has pK_a values of 1.27 and 4.27. Therefore, the buffer regions will be centered at about pH 1.3 and pH 4.3. The buffer regions carry the information necessary to get the pK_a values as the concentrations of acid and conjugate base change along a buffer region.

Between the two buffer regions there is an end-point, or equivalence point, at about pH 3. This end-point is not sharp and is typical of a diprotic acid whose buffer regions overlap by a small amount: pK_{a2} – pK_{a1} is about three in this example. (If the difference in pK values were about two or less, the end-point would not be noticeable.) The second end-point begins at about pH 6.3 and is sharp. This indicates that all the protons have been removed. When this is so, the solution is not buffered and the pH rises steeply on addition of a small amount of strong base. However, the pH does not continue to rise indefinitely. A new buffer region begins at about pH 11 (pK_w – 3), which is where self-ionization of water becomes important.

It is very difficult to measure pH values of less than two in aqueous solution with a glass electrode, because the Nernst equation breaks down at such low pH values. To determine pK values of less than about 2 or more than about 11 spectrophotometric or NMR measurements may be used instead of, or combined with, pH measurements.

When the glass electrode cannot be employed, as with non-aqueous solutions, spectrophotometric methods are frequently used. These may involve absorbance or fluorescence measurements. In both cases the measured quantity is assumed to be proportional to the sum of contributions from each photo-active species; with absorbance measurements the Beer-Lambert law is assumed to apply.

Aqueous solutions with normal water cannot be used for ^{1}H NMR measurements but heavy water, D_2O, must be used instead. ^{13}C NMR data, however, can be used with normal water and ^{1}H NMR spectra can be used with non-aqueous media. The quantities measured with NMR are time-averaged chemical shifts, as proton exchange is fast on the NMR time-scale. Other chemical shifts, such as those of ^{31}P can be measured.

Micro-constants

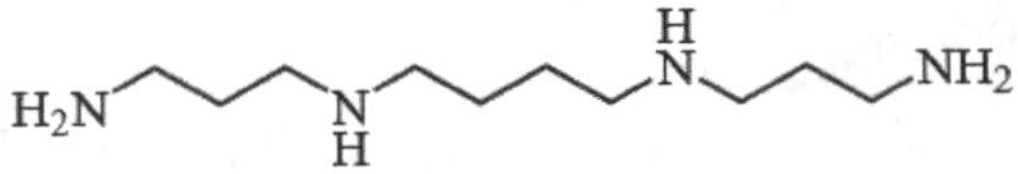

spermine

A base such as spermine has a few different sites where protonation can occur. In this example the first proton can go on the terminal $-NH_2$ group, or either of the internal $-NH-$ groups. The pK_a values for dissociation of spermine protonated at one or other of the sites are examples of micro-constants. They cannot be determined directly by means of pH, absorbance, fluorescence or NMR measurements. Nevertheless, the site of protonation is very important for biological function, so mathematical methods have been developed for the determination of micro-constants.

Applications and Significance

A knowledge of pK_a values is important for the quantitative treatment of systems involving acid–base equilibria in solution. Many applications exist in biochemistry; for example, the pK_a values of proteins and amino acid side chains are of major importance for the activity of enzymes and the stability of proteins. Protein pK_a values cannot always be measured directly, but may be calculated using theoretical methods. Buffer solutions are used extensively to provide solutions at or near the physiological pH for the study of biochemical reactions; the design of these solutions depends on a knowledge of the pK_a values of their components. Important buffer solutions include MOPS, which provides a solution with pH 7.2, and tricine, which is used in gel electrophoresis. Buffering is an essential part of acid base physiology including acid–base homeostasis, and is key to understanding disorders such as acid–base imbalance. The isoelectric point of a given molecule is a function of its pK values, so different molecules have different isoelectric points. This permits a technique called isoelectric focusing, which is used for separation of proteins by 2-D gel polyacrylamide gel electrophoresis.

Buffer solutions also play a key role in analytical chemistry. They are used whenever there is a need to fix the pH of a solution at a particular value. Compared with an aqueous solution, the pH of a buffer solution is relatively insensitive to the addition of a small amount of strong acid or strong base. The buffer capacity of a simple buffer solution is largest when pH = pK_a. In acid–base extraction, the efficiency of extraction of a compound into an organic phase, such as an ether, can be optimised by adjusting the pH of the aqueous phase using an appropriate buffer. At the optimum pH, the concentration of the electrically neutral species is maximised; such a species is more soluble in organic solvents having a low dielectric constant than it is in water. This technique is used for the purification of weak acids and bases.

A pH indicator is a weak acid or weak base that changes colour in the transition pH range, which is approximately pK_a ± 1. The design of a universal indicator requires a mixture of indicators whose adjacent pK_a values differ by about two, so that their transition pH ranges just overlap.

In pharmacology ionization of a compound alters its physical behaviour and macro properties such as solubility and lipophilicity (log *p*). For example, ionization of any compound will increase the solubility in water, but decrease the lipophilicity. This is exploited in drug development to increase the concentration of a compound in the blood by adjusting the pK_a of an ionizable group.

Knowledge of pK_a values is important for the understanding of coordination complexes, which are formed by the interaction of a metal ion, M^{m+}, acting as a Lewis acid, with a ligand, L, acting as a Lewis base. However, the ligand may also undergo protonation reactions, so the formation of a complex in aqueous solution could be represented symbolically by the reaction,

$$[M(H_2O)_n]^{m+} + LH \rightleftharpoons [M(H_2O)_{n-1}L]^{(m-1)+} + H_3O^+$$

To determine the equilibrium constant for this reaction, in which the ligand loses a proton, the pK_a of the protonated ligand must be known. In practice, the ligand may be polyprotic; for example $EDTA^{4-}$ can accept four protons; in that case, all pK_a values must be known. In addition, the metal ion is subject to hydrolysis, that is, it behaves as a weak acid, so the pK values for the hydrolysis reactions must also be known. Assessing the hazard associated with an acid or base may require a knowledge of pK_a values. For example, hydrogen cyanide is a very toxic gas, because the cyanide ion inhibits the iron-containing enzyme cytochrome c oxidase. Hydrogen cyanide is a weak acid in aqueous solution with a pK_a of about 9. In strongly alkaline solutions, above pH 11, say, it follows that sodium cyanide is "fully dissociated" so the hazard due to the hydrogen cyanide gas is much reduced. An acidic solution, on the other hand, is very hazardous because all the cyanide is in its acid form. Ingestion of cyanide by mouth is potentially fatal, independently of pH, because of the reaction with cytochrome c oxidase.

In environmental science acid–base equilibria are important for lakes and rivers; for example, humic acids are important components of natural waters. Another example occurs in chemical oceanography: in order to quantify the solubility of iron(III) in seawater at various salinities, the pK_a values for the formation of the iron(III) hydrolysis products $Fe(OH)^{2+}$, $Fe(OH)_2^+$ and $Fe(OH)_3$ were determined, along with the solubility product of iron hydroxide.

Values for Common Substances

There are multiple techniques to determine the pK_a of a chemical, leading to some discrepancies between different sources. Well measured values are typically within 0.1 units of each other. Data presented here were taken at 25 °C in water. More values can be found in thermodynamics.

Chemical Name	Equilibrium	pK_a
BH = Adenine	$BH \rightleftharpoons B^- + H^+$	4.17
	$BH+2 \rightleftharpoons BH + H^+$	9.65
H_3A = Arsenic acid	$H_3A \rightleftharpoons H_2A^- + H^+$	2.22
	$H_2A^- \rightleftharpoons HA^{2-} + H^+$	6.98
	$HA^{2-} \rightleftharpoons A^{3-} + H^+$	11.53
HA = Benzoic acid	$HA \rightleftharpoons H^+ + A^-$	4.204
HA = Butyric acid	$HA \rightleftharpoons H^+ + A^-$	4.82
H_2A = Chromic acid	$H_2A \rightleftharpoons HA^- + H^+$	0.98
	$HA^- \rightleftharpoons A^{2-} + H^+$	6.5
B = Codeine	$BH^+ \rightleftharpoons B + H^+$	8.17
HA = Cresol	$HA \rightleftharpoons H^+ + A^-$	10.29
HA = Formic acid	$HA \rightleftharpoons H^+ + A^-$	3.751
HA = Hydrofluoric acid	$HA \rightleftharpoons H^+ + A^-$	3.17
HA = Hydrocyanic acid	$HA \rightleftharpoons H^+ + A^-$	9.21
HA = Hydrogen selenide	$HA \rightleftharpoons H^+ + A^-$	3.89
HA = Hydrogen peroxide (90%)	$HA \rightleftharpoons H^+ + A^-$	11.7
HA = Lactic acid	$HA \rightleftharpoons H^+ + A^-$	3.86
HA = Propionic acid	$HA \rightleftharpoons H^+ + A^-$	4.87
HA = Phenol	$HA \rightleftharpoons H^+ + A^-$	9.99
H_2A = L-(+)-Ascorbic Acid	$H_2A \rightleftharpoons HA^- + H^+$	4.17
	$HA^- \rightleftharpoons A^{2-} + H^+$	11.57

pH

In chemistry, pH (potential of hydrogen) is a numeric scale used to specify the acidity or basicity of an aqueous solution. It is approximately the negative of the base 10 logarithm of the molar concentration, measured in units of moles per liter, of hydrogen ions. More precisely it is the negative of the logarithm to base 10 of the activity of the hydrogen ion. Solutions with a pH less than 7 are acidic and solutions with a pH greater than 7 are basic. Pure water is neutral, at pH 7, being neither an acid nor a base. Contrary to popular belief, the pH value can be less than 0 or greater than 14 for very strong acids and bases respectively.

Lemon juice tastes sour because it contains 5% to 6% citric acid and has a pH of 2.2. (high acidity).

pH measurements are important in agronomy, medicine, biology, chemistry, agriculture, forestry, food science, environmental science, oceanography, civil engineering, chemical engineering, nutrition, water treatment and water purification, as well as many other applications.

The pH scale is traceable to a set of standard solutions whose pH is established by international agreement. Primary pH standard values are determined using a concentration cell with transference, by measuring the potential difference between a hydrogen electrode and a standard electrode such as the silver chloride electrode. The pH of aqueous solutions can be measured with a glass electrode and a pH meter, or an indicator.

History

The concept of pH was first introduced by the Danish chemist Søren Peder Lauritz Sørensen at the Carlsberg Laboratory in 1909 and revised to the modern pH in 1924 to accommodate definitions and measurements in terms of electrochemical cells. In the first papers, the notation had the “H” as a subscript to the lowercase “p”, as so: p_H.

The exact meaning of the “p” in “pH” is disputed, but according to the Carlsberg Foundation, pH stands for “power of hydrogen”. It has also been suggested that the “p” stands for the German *Potenz* (meaning “power”), others refer to French *puissance* (also meaning “power”, based on the fact that the Carlsberg Laboratory was French-speaking). Another suggestion is that the “p” stands for the Latin terms *pondus hydrogenii* (quantity of hydrogen), *potentia hydrogenii* (capacity of hydrogen), or potential hydrogen. It is also suggested that Sørensen used the letters “p” and “q” (commonly paired letters in mathematics) simply to label the test solution (p) and the reference solution (q). Currently in chemistry, the p stands for “decimal cologarithm of”, and is also used in the term pK_a, used for acid dissociation constants.

Definition and Measurement

pH

pH is defined as the decimal logarithm of the reciprocal of the hydrogen ion activity, a_{H^+}, in a solution.

$$\mathrm{pH} = -\log_{10}(a_{\mathrm{H}^+}) = \log_{10}\left(\frac{1}{a_{\mathrm{H}^+}}\right)$$

For example, a solution with a hydrogen ion activity of $5\times10^{-6} = 1/(2\times10^5)$ (at that level essentially the number of moles of hydrogen ions per liter of solution) has a pH of $\log_{10}(2\times10^5) = 5.3$. For a commonplace example based on the facts that the masses of a mole of water, a mole of hydrogen ions, and a mole of hydroxide ions are respectively 18 g, 1 g, and 17 g, a quantity of 10^7 moles of pure (pH 7) water, or 180 tonnes (18×10^7 g), contains close to 1 g of dissociated hydrogen ions (or rather 19 g of H_3O^+ hydronium ions) and 17 g of hydroxide ions.

This definition was adopted because ion-selective electrodes, which are used to measure pH, respond to activity. Ideally, electrode potential, *E*, follows the Nernst equation, which, for the hydrogen ion can be written as,

$$E = E^0 + \frac{RT}{F}\ln(a_{\mathrm{H}^+}) = E^0 - \frac{2.303RT}{F}\mathrm{pH}$$

where *E* is a measured potential, E^o is the standard electrode potential, *R* is the gas constant, *T* is the temperature in kelvins, *F* is the Faraday constant. For H^+ number of electrons transferred is one. It follows that electrode potential is proportional to pH when pH is defined in terms of activity. Precise measurement of pH is presented in International Standard ISO 31-8 as follows: A galvanic cell is set up to measure the electromotive force (e.m.f.) between a reference electrode and an electrode sensitive to the hydrogen ion activity when they are both immersed in the same aqueous solution. The reference electrode may be a silver chloride electrode or a calomel electrode. The hydrogen-ion selective electrode is a standard hydrogen electrode.

Reference electrode | concentrated solution of KCl || test solution | H_2 | Pt

Firstly, the cell is filled with a solution of known hydrogen ion activity and the emf, E_S, is measured. Then the emf, E_X, of the same cell containing the solution of unknown pH is measured.

$$\mathrm{pH(X)} = \mathrm{pH(S)} + \frac{E_S - E_X}{z}$$

The difference between the two measured emf values is proportional to pH. This method of calibration avoids the need to know the standard electrode potential. The proportionality constant, $1/z$ is ideally equal to $\frac{1}{2.303RT/F}$ the "Nernstian slope".

To apply this process in practice, a glass electrode is used rather than the cumbersome hydrogen electrode. A combined glass electrode has an in-built reference electrode. It is calibrated against buffer solutions of known hydrogen ion activity. IUPAC has proposed the use of a set of buffer solutions of known H^+ activity. Two or more buffer solutions are used in order to accommodate the fact that the "slope" may differ slightly from ideal. To implement this approach to calibration, the electrode is first immersed in a standard solution and the reading on a pH meter is adjusted to be equal to the standard buffer's value. The reading from a second standard buffer solution is then adjusted, using the "slope" control, to be equal to the pH for that solution. Further details, are given in the IUPAC recommendations. When more than two buffer solutions are used the electrode is calibrated by fitting observed pH values to a straight line with respect to standard buffer values. Commercial standard buffer solutions usually come with information on the value at 25 °C and a correction factor to be applied for other temperatures.

The pH scale is logarithmic and therefore pH is a dimensionless quantity.

p[H]

This was the original definition of Sørensen, which was superseded in favor of pH in 1909. However, it is possible to measure the concentration of hydrogen ions directly, if the electrode is calibrated in terms of hydrogen ion concentrations. One way to do this, which has been used extensively, is to titrate a solution of known concentration of a strong acid with a solution of known concentration of strong alkaline in the presence of a relatively high concentration of background electrolyte. Since the concentrations of acid and alkaline are known, it is easy to calculate the concentration of hydrogen ions so that the measured potential can be correlated with concentrations. The calibration is usually carried out using a Gran plot. The calibration yields a value for the standard electrode potential, E^o, and a slope factor, f, so that the Nernst equation in the form,

$$E = E^0 + f\frac{2.303RT}{F}\log[H^+]$$

can be used to derive hydrogen ion concentrations from experimental measurements of E. The slope factor, f, is usually slightly less than one. A slope factor of less than 0.95 indicates that the electrode is not functioning correctly. The presence of background electrolyte ensures that the hydrogen ion activity coefficient is effectively constant during the titration. As it is constant, its value can be set to one by defining the standard state as being the solution containing the background electrolyte. Thus, the effect of using this procedure is to make activity equal to the numerical value of concentration.

The glass electrode (and other ion selective electrodes) should be calibrated in a medium similar to the one being investigated. For instance, if one wishes to measure the pH of a seawater sample, the electrode should be calibrated in a solution resembling seawater in its chemical composition, as detailed below.

The difference between p[H] and pH is quite small. It has been stated that pH = p[H] + 0.04. It is common practice to use the term "pH" for both types of measurement.

pH Indicators

Indicators may be used to measure pH, by making use of the fact that their color changes with pH. Visual comparison of the color of a test solution with a standard color chart provides a means to measure pH accurate to the nearest whole number. More precise measurements are possible if the color is measured spectrophotometrically, using a colorimeter or spectrophotometer. Universal indicator consists of a mixture of indicators such that there is a continuous color change from about pH 2 to pH 10. Universal indicator paper is made from absorbent paper that has been impregnated with universal indicator. Another method of measuring pH is using an electronic pH meter.

Chart showing the variation of color of universal indicator paper with pH.

pOH

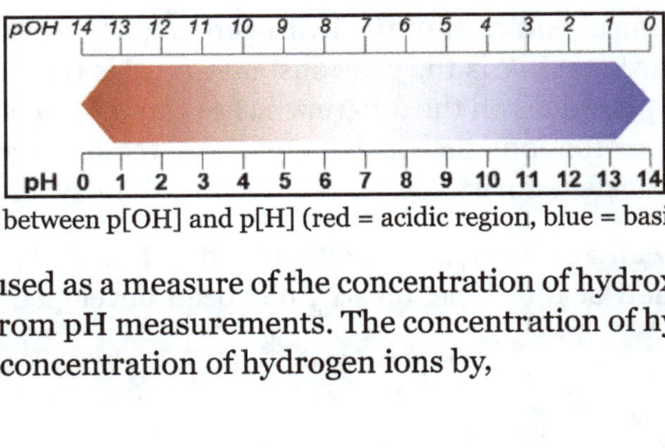

Relation between p[OH] and p[H] (red = acidic region, blue = basic region).

pOH is sometimes used as a measure of the concentration of hydroxide ions. OH^-. pOH values are derived from pH measurements. The concentration of hydroxide ions in water is related to the concentration of hydrogen ions by,

$$[OH^-] = \frac{K_W}{[H^+]}$$

where K_w is the self-ionisation constant of water. Taking logarithms,

$$pOH = pK_W - pH$$

So, at room temperature, pOH ≈ 14 – pH. However this relationship is not strictly valid in other circumstances, such as in measurements of soil alkalinity.

Extremes of pH

Measurement of pH below about 2.5 (ca. 0.003 mol dm^{-3} acid) and above about 10.5 (ca. 0.0003 mol dm^{-3} alkaline) requires special procedures because, when using the glass electrode, the Nernst law breaks down under those conditions. Various factors contribute to this. It cannot be assumed that liquid junction potentials are independent of pH. Also, extreme pH implies that the solution is concentrated, so electrode potentials are affected by ionic strength variation. At high pH the glass electrode may be affected by "alkaline error", because the electrode becomes sensitive to the concentration of cations such as Na^+ and K^+ in the solution. Specially constructed electrodes are available which partly overcome these problems.

Runoff from mines or mine tailings can produce some very low pH values.

Non-aqueous Solutions

Hydrogen ion concentrations (activities) can be measured in non-aqueous solvents. pH values based on these measurements belong to a different scale from aqueous pH values, because activities relate to different standard states. Hydrogen ion activity, a_{H^+}, can be defined as:

$$a_{H^+} = \exp\left(\frac{\mu_{H^+} - \mu^{\ominus}_{H^+}}{RT}\right)$$

where μ_{H^+} is the chemical potential of the hydrogen ion, $\mu^{o}_{H^+}$ is its chemical potential in the chosen standard state, R is the gas constant and T is the thermodynamic temperature. Therefore, pH values on the different scales cannot be compared directly due to different solvated proton ions such as lyonium ions, requiring an intersolvent scale which involves the transfer activity coefficient of hydronium/lyonium ion.

pH is an example of an acidity function. Other acidity functions can be defined. For example, the Hammett acidity function, H_0, has been developed in connection with superacids.

The concept of "Unified pH scale" has been developed on the basis of the absolute chemical potential of the proton. This scale applies to liquids, gases and even solids.

Applications

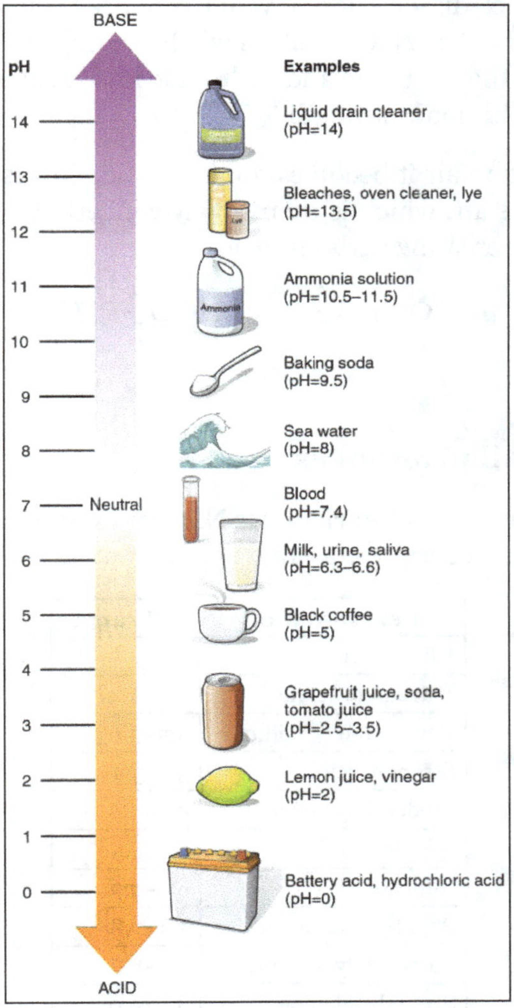

pH values of some common substances.

Pure water is neutral. When an acid is dissolved in water, the pH will be less than 7. When a base, or alkali, is dissolved in water, the pH will be greater than 7. A solution of a strong acid, such as hydrochloric acid, at concentration 1 mol dm^{-3} has a pH of 0. A solution of a strong alkali, such as sodium hydroxide, at concentration 1 mol dm^{-3}, has a pH of 14. Thus, measured pH values will lie mostly in the range 0 to 14, though negative pH values and values above 14 are entirely possible. Since pH is a logarithmic scale, a difference of one pH unit is equivalent to a tenfold difference in hydrogen ion concentration.The pH of neutrality is not exactly 7, although this is a good approximation in most cases. Neutrality is defined as the condition where $[H^+] = [OH^-]$ (or the activities are equal). Since self-ionization of water holds the product of these concentration $[H^+]\times[OH^-] = Kw$, it can be seen that at neutrality $[H^+] = [OH^-] = sqrt(Kw)$, or $pH = pKw/2$. pKw is approximately 14 but depends on ionic strength and temperature, and so the pH of neutrality does also. Pure water and a solution of NaCl in pure water are both neutral, since dissociation of water produces equal numbers of both ions. However the pH of the neutral NaCl solution will be slightly different from that of neutral pure water because the hydrogen and hydroxide ions' activity is dependent on ionic strength, so K_w varies with ionic strength.

If pure water is exposed to air it becomes mildly acidic. This is because water absorbs carbon dioxide from the air, which is then slowly converted into bicarbonate and hydrogen ions (essentially creating carbonic acid).

$$CO_2 + H_2O <=> HCO_3^- + H^+$$

pH in Soil

Classification of Soil pH Ranges

The United States Department of Agriculture Natural Resources Conservation Service, formerly Soil Conservation Service classifies soil pH ranges as follows:

Denomination	pH range
Ultra acidic	< 3.5
Extremely acidic	3.5–4.4
Very strongly acidic	4.5–5.0
Strongly acidic	5.1–5.5
Moderately acidic	5.6–6.0
Slightly acidic	6.1–6.5
Neutral	6.6–7.3
Slightly alkaline	7.4–7.8
Moderately alkaline	7.9–8.4
Strongly alkaline	8.5–9.0
Very strongly alkaline	> 9.0

pH in Nature

pH-dependent plant pigments that can be used as pH indicators occur in many plants, including hibiscus, red cabbage (anthocyanin) and red wine. The juice of citrus fruits is acidic mainly because it contains citric acid. Other carboxylic acids occur in many living systems. For example, lactic acid is produced by muscle activity. The state of protonation of phosphate derivatives, such as ATP, is pH-dependent. The functioning of the oxygen-transport enzyme hemoglobin is affected by pH in a process known as the Root effect.

Seawater

The pH of seawater is typically limited to a range between 7.5 and 8.4. It plays an important role in the ocean's carbon cycle, and there is evidence of ongoing ocean acidification caused by carbon dioxide emissions. However, pH measurement is complicated by the chemical properties of seawater, and several distinct pH scales exist in chemical oceanography.

As part of its operational definition of the pH scale, the IUPAC defines a series of buffer solutions across a range of pH values (often denoted with NBS or NIST designation). These solutions have a relatively low ionic strength (~0.1) compared to that of seawater (~0.7), and, as a consequence, are not recommended for use in characterizing the pH of seawater, since the ionic strength differences cause changes in electrode potential. To resolve this problem, an alternative series of buffers based on artificial seawater was developed. This new series resolves the problem of ionic strength differences between samples and the buffers, and the new pH scale is referred to as the 'total scale', often denoted as pH_T. The total scale was defined using a medium containing sulfate ions. These ions experience protonation, $H^+ + SO_4^{2-} \rightleftharpoons HSO_4^-$, such that the total scale includes the effect of both protons (free hydrogen ions) and hydrogen sulfate ions:

$$[H^+]_T = [H^+]_F + [HSO_4^-]$$

An alternative scale, the 'free scale', often denoted 'pH_F', omits this consideration and focuses solely on $[H^+]_F$, in principle making it a simpler representation of hydrogen ion concentration. Only $[H^+]_T$ can be determined, therefore $[H^+]_F$ must be estimated using the [SO2−4] and the stability constant of HSO−4, K* S:

$$[H^+]_F = [H^+]_T - [HSO_4^-] = [H^+]_T \left(1 + [SO_4^{2-}] / K_S^*\right)^{-1}$$

However, it is difficult to estimate K*S in seawater, limiting the utility of the otherwise more straightforward free scale.

Another scale, known as the 'seawater scale', often denoted 'pH_{SWS}', takes account of a further protonation relationship between hydrogen ions and fluoride ions, $H^+ + F^- \rightleftharpoons HF$. Resulting in the following expression for $[H^+]_{SWS}$:

$$[H^+]_{SWS} = [H^+]_F + [HSO_4^-] + [HF]$$

However, the advantage of considering this additional complexity is dependent upon the abundance of fluoride in the medium. In seawater, for instance, sulfate ions occur at much greater concentrations (>400 times) than those of fluoride. As a consequence, for most practical purposes, the difference between the total and seawater scales is very small.

The following three equations summarise the three scales of pH:

$$pH_F = -\log\left[H^+\right]_F$$
$$pH_T = -\log\left(\left[H^+\right]_F + [HSO_4^-]\right) = -\log\left[H^+\right]_T$$
$$pH_{SWS} = -\log\left(\left[H^+\right]_F + [HSO_4^-] + [HF]\right) = -\log\left[H^+\right]_{SWS}$$

In practical terms, the three seawater pH scales differ in their values by up to 0.12 pH units, differences that are much larger than the accuracy of pH measurements typically required, in particular, in relation to the ocean's carbonate system. Since it omits consideration of sulfate and fluoride ions, the free scale is significantly different from both the total and seawater scales. Because of the relative unimportance of the fluoride ion, the total and seawater scales differ only very slightly.

Living Systems

pH in living systems	
Compartment	**pH**
Gastric acid	1.5-3.5
Lysosomes	4.5
Human skin	4.7
Granules of chromaffin cells	5.5
Urine	6.0
Cytosol	7.2
Blood (natural pH)	7.34–7.45
Cerebrospinal fluid (CSF)	7.5
Mitochondrial matrix	7.5
Pancreas secretions	8.1

The pH of different cellular compartments, body fluids, and organs is usually tightly regulated in a process called acid-base homeostasis. The most common disorder in acid-base homeostasis is acidosis, which means an acid overload in the body, generally defined by pH falling below 7.35. Alkalosis is the opposite condition, with blood pH being excessively high.

The pH of blood is usually slightly basic with a value of pH 7.365. This value is often referred to as physiological pH in biology and medicine. Plaque can create a local acidic environment that can result in tooth decay by demineralization. Enzymes and other proteins have an optimum pH range and can become inactivated or denatured outside this range.

Calculations of pH

The calculation of the pH of a solution containing acids and/or bases is an example of a chemical speciation calculation, that is, a mathematical procedure for calculating the concentrations of all chemical species that are present in the solution. The complexity of the procedure depends on the nature of the solution. For strong acids and bases no calculations are necessary except in extreme situations. The pH of a solution containing a weak acid requires the solution of a quadratic equation. The pH of a solution containing a weak base may require the solution of a cubic equation. The general case requires the solution of a set of non-linear simultaneous equations.

A complicating factor is that water itself is a weak acid and a weak base. It dissociates according to the equilibrium,

$$2H_2O <=> H_3O^+(aq) + OH^-(aq)$$

with a dissociation constant, K_w defined as,

$$K_w = [H^+][OH^-]$$

where $[H^+]$ stands for the concentration of the aqueous hydronium ion and $[OH^-]$ represents the concentration of the hydroxide ion. This equilibrium needs to be taken into account at high pH and when the solute concentration is extremely low.

Strong Acids and Bases

Strong acids and bases are compounds that, for practical purposes, are completely dissociated in water. Under normal circumstances this means that the concentration of hydrogen ions in acidic solution can be taken to be equal to the concentration of the acid. The pH is then equal to minus the logarithm of the concentration value. Hydrochloric acid (HCl) is an example of a strong acid. The pH of a 0.01M solution of HCl is equal to $-\log_{10}(0.01)$, that is, pH = 2. Sodium hydroxide, NaOH, is an example of a strong base. The p[OH] value of a 0.01M solution of NaOH is equal to $-\log_{10}(0.01)$, that is, p[OH] = 2. From the definition of p[OH] above, this means that the pH is equal to about 12. For solutions of sodium hydroxide at higher concentrations the self-ionization equilibrium must be taken into account.

Self-ionization must also be considered when concentrations are extremely low. Consider, for example, a solution of hydrochloric acid at a concentration of 5×10^{-8}M. The simple procedure given above would suggest that it has a pH of 7.3. This is clearly wrong as an acid solution should have a pH of less than 7. Treating the system as a mixture of hydrochloric acid and the amphoteric substance water, a pH of 6.89 results.

Weak Acids and Bases

A weak acid or the conjugate acid of a weak base can be treated using the same formalism.

$$\text{Acid}: HA \rightleftharpoons H^+ + A^-$$

$$\text{Base}: HA^+ \rightleftharpoons H^+ \quad A$$

First, an acid dissociation constant is defined as follows. Electrical charges are omitted from subsequent equations for the sake of generality,

$$K_a = \frac{[H][A]}{[HA]}$$

and its value is assumed to have been determined by experiment. This being so, there are three unknown concentrations, [HA], [H^+] and [A^-] to determine by calculation. Two additional equations are needed. One way to provide them is to apply the law of mass conservation in terms of the two "reagents" H and A.

$$C_A = [A] + [HA]$$

$$C_H = [H] + [HA]$$

C stands for analytical concentration. In some texts one mass balance equation is replaced by an equation of charge balance. This is satisfactory for simple cases like this one, but is more difficult to apply to more complicated cases as those below. Together with the equation defining K_a, there are now three equations in three unknowns. When an acid is dissolved in water $C_A = C_H = C_a$, the concentration of the acid, so [A] = [H]. After some further algebraic manipulation an equation in the hydrogen ion concentration may be obtained.

$$[H]^2 + K_a[H] - K_aC_a = 0$$

Solution of this quadratic equation gives the hydrogen ion concentration and hence p[H] or, more loosely, pH. This procedure is illustrated in an ICE table which can also be used to calculate the pH when some additional (strong) acid or alkaline has been added to the system, that is, when $C_A \neq C_H$.

For example, what is the pH of a 0.01M solution of benzoic acid, $pK_a = 4.19$?

- Step 1: $K_a = 10^{-4.19} = 6.46 \times 10^{-5}$
- Step 2: Set up the quadratic equation. $[H]^2 + 6.46 \times 10^{-5}[H] - 6.46 \times 10^{-7} = 0$
- Step 3: Solve the quadratic equation. $[H^+] = 7 \cdot 74 \times 10^{-4}$; pH = 3.11

For alkaline solutions an additional term is added to the mass-balance equation for hydrogen. Since addition of hydroxide reduces the hydrogen ion concentration, and the hydroxide ion concentration is constrained by the self-ionization equilibrium to be equal to $\frac{K_w}{[H+]}$

$$C_H = \frac{[H]+[HA]-K_w}{[H]}$$

In this case the resulting equation in [H] is a cubic equation.

General Method

Some systems, such as with polyprotic acids, are amenable to spreadsheet calculations. With three or more reagents or when many complexes are formed with general formulae such as $A_pB_qH_r$ the following general method can be used to calculate the pH of a solution. For example, with three reagents, each equilibrium is characterized by an equilibrium constant, β.

$$[A_pB_qH_r] = \beta_{pqr}[A]^p[B]^q[H]^r$$

Next, write down the mass-balance equations for each reagent:

$$C_A = [A] + \Sigma p\beta_{pqr}[A]^p[B]^q[H]^r$$
$$C_B = [B] + \Sigma q\beta_{pqr}[A]^p[B]^q[H]^r$$
$$C_H = [H] + \Sigma r\beta_{pqr}[A]^p[B]^q[H]^r - K_w[H]^{-1}$$

Note that there are no approximations involved in these equations, except that each stability constant is defined as a quotient of concentrations, not activities. Much more complicated expressions are required if activities are to be used.

There are 3 non-linear simultaneous equations in the three unknowns, [A], [B] and [H]. Because the equations are non-linear, and because concentrations may range over many powers of 10, the solution of these equations is not straightforward. However, many computer programs are available which can be used to perform these calculations. There may be more than three reagents. The calculation of hydrogen ion concentrations, using this formalism, is a key element in the determination of equilibrium constants by potentiometric titration.

Concepts

A compound is classified as an acid or a base based on certain properties. At present there are several theories which define the concepts of acidity and basicity. Some important concepts are detailed below.

Arrhenius concept of acids and bases : The Arrhenius definition of acids says that they are compounds that give off H^+ ions in water and that bases are compounds that give off OH^- ions in water. Thus, according to this theory only protic acids are allowed and only hydroxide bases are allowed to be classified as an acid or a base.

$$H_2SO_4\,(aq) \rightarrow H^+\,(aq) + HSO_4^-\,(aq)$$

Bronsted-Lowry definition : The Bronsted-Lowry definition of acids is that acids are compounds that give off H^+ ions when they react with another compound. Likewise, this definition says that bases are compounds that accept H^+ ions from other compounds. Furthermore, it also brings a new concept of conjugate acids and conjugate bases. Since an acid after donating its proton is technically a base according to this definition and is referred to as a conjugate acid. So every acid has its conjugate base and vice versa. Also, the stronger is an acid, the weaker is its conjugate base and vice versa.

$$\underset{\text{Acid}}{H_2SO_4} + \underset{\text{Base}}{H_2O} \rightarrow \underset{\text{Conjugate acid}}{H_3O^+} + \underset{\text{Conjugate base}}{HSO_4^-}$$

Lewis concept of acids and bases : Acids are electron pair acceptors while bases are electron pair donors. Thus, electron deficient species like BF_3 are Lewis acids while electron rich species such as tertiary amines are Lewis bases. Lewis acids may combine with Lewis bases to generate a salt.

$$\underset{\text{a Lewis base}}{(CH_3)_3N:} + \underset{\text{a Lewis acid}}{BF_3} \rightarrow (CH_3)_3\overset{+}{N}-\overset{-}{B}F_3$$

pH and pKa

The strength of an acid depends on its degree of dissociation in water and can be determined from the equilibrium expression.

$$HA + H_2O \rightleftharpoons H_3O^+ + A^-$$

$$K_a = \frac{[H_3O^+][A^-]}{[HA]}$$

$$-\log K_a = -\log[H^+] - \log\frac{|A^-|}{|HA|}$$

$$\text{or} \quad pK_a = pH - \log\frac{|A^-|}{|HA|}$$

$$pH = pK_a + \log\frac{|A^-|}{|HA|}$$

This equation is known as Hendersson-Hasselbach equation and relates a relationship between the concentration of hydrogen ions in solutions and acid strength. Thus, pH is a measurement of the H^+ concentration in a liquid. If there is a high H^+ concentration, the pH indicates that it is a very acidic solution. If the solution is neutral, there is only a small H^+ concentration, and the pH reflects that. If the solution is basic, there is almost no H^+ concentration, and you can tell that by the pH number.

As with acids, a similar scale may be derived for bases based on the equilibrium of the protonated form of the base and base itself.

$$B: + H_2O \rightleftharpoons BH^+ + OH^-$$

$$K_b = \frac{[BH^+][OH]}{[B:]}$$

However, such an expression creates two separate scales for the determination of strengths of acids and bases. Instead the strength of the conjugate acid is determined then a common scale in terms of pK_a may be used for both acids and bases. Thus, the lower the value of pK_a, stronger is the acid. Similarly, higher is the value of pK_a of base, more basic it is.

Buffer Solution

A solution of a weak acid and its conjugate base is referred as a buffer solution. A buffer solution has the ability to resist to a small change of pH, because the weak acid can donate a proton to any HO^- added to the solution, and its conjugate base can accept any H^+, which is added to the solution.

Can donate an H^+ to HO^-

$$HA + HO^- \rightarrow A^- + H_2O$$

$$A^- + H_3O^+ \rightarrow HA + H_2O$$

Can donate an H^+ from H_3O^+

A common example of buffer is a solution of sodium acetate in acetic acid. Human blood is also a buffered solution and maintains a pH close to neutrality.

The Effect of Structure on pKa

There are several factors which decide the relative strength of an acid (H-A). The strength of the H-A bond, the electronegativity of A^- and the factors stabilizing A^- with respect to HA are some of these factors. The first one is usually not much relevant but may be the cause of difference in acid strengths of CH_3 OH (16) vs CH_4 (43).

- The stabilization of the anion with respect to the undissociated molecule is however of quite importance. As an example, in case of formic acid, the anion generated is greatly stabilized due to effective delocalization of the charge into the two canonical structures of identical energy. Such delocalization is not possible in the formic acid itself without leading to separation of charges.

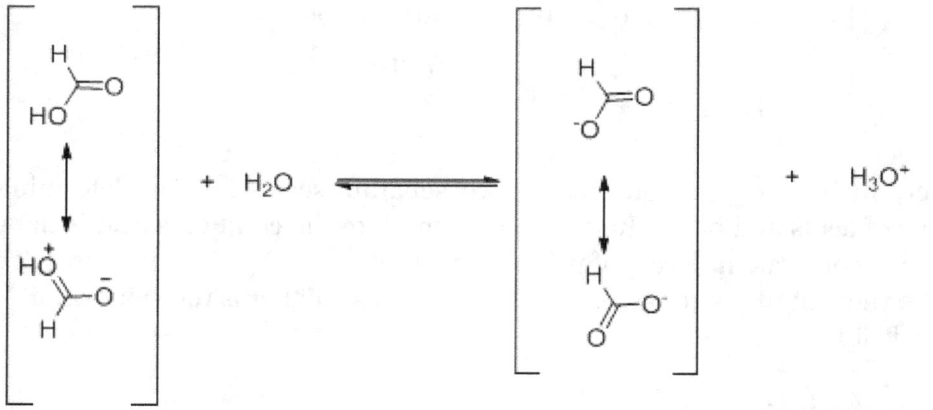

- Similarly, the solvent plays an important role. The solvation of the species generated in the solvent could lead to an increase or decrease in acidity. As an acid or base is dissociated, more charges develop which may be better stabilized in solvent having a high dielectric constant and a greater degree of hydrogen bonding ability. Thus, HCl when dissolves in water is very strong acid as the ions it generates are suitably solvated by the water molecules but if the solvent is changed to toluene, there is hardly any dissociation.

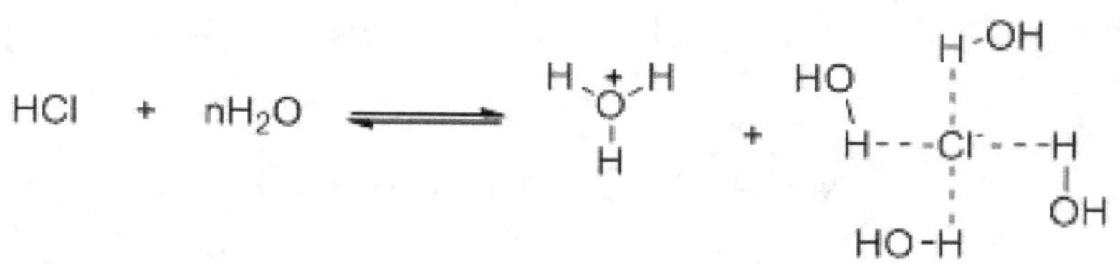

- It is also expected that electron withdrawing and electron donating groups will have a profound effect on the acid strength of a substance. Thus, electron donating group substituted carboxylic acids are less acidic than those without them. In comparison of the acid strengths of methanoic and ethanoic acids, it is observed that the former has pKa of 3.77 while for the later has only 4.84. The difference has been explained by the electron donating inductive effects of the methyl group of ethanoic acid. However, it is to be noted that the inductive effect falls off quickly with the increase in the chain length and other factors become dominant on going down the series.

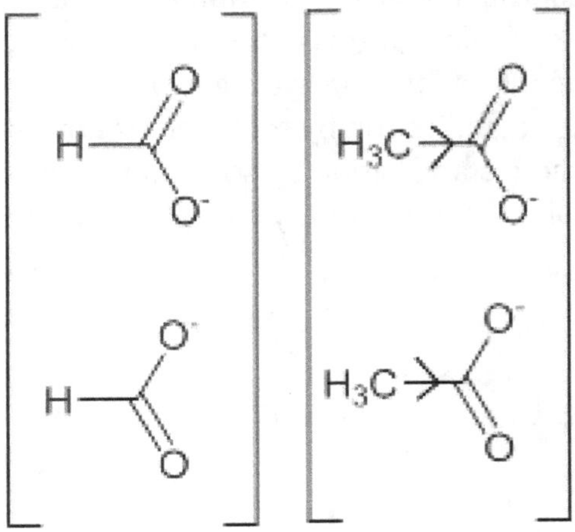

- Similarly, substitutions by electron withdrawing groups tend to increase the acidity of an acid. Thus, in the following series of halo acids it is observed that the pK_a decreases with increasing number of halogen atoms. Furthermore, it is also observed that among the monosubstituted halo acids the value of pK_a decreases with increase in electronegativity of the halogen. This indicates that the more is the electron withdrawing power of a group, the lower is the pK_a.

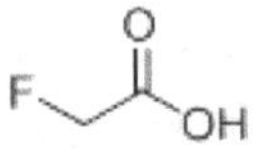

2.57

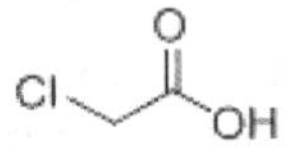

2.86

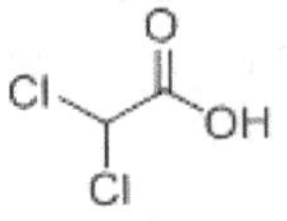

1.25

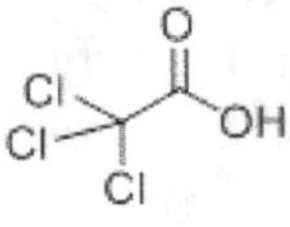

0.65

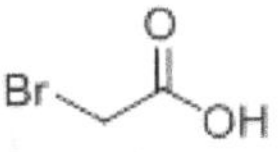

2.90

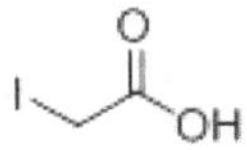

3.16

- In case of aliphatic bases, it is thus expected that the increase in the number of alkyl groups will increase the basicity. Thus, basicity is supposed to increase on going from primary to secondary to tertiary amines. In fact, it is observed that in a solution of water, the basicity does increase on going from the primary amine to secondary amine but drops slightly on going to the tertiary amine. This drop is explained by considering the factor of stabilization of the conjugate acid by solvation. As the number of alkyl groups increase, there is a decrease in the number of hydrogen atoms capable of forming hydrogen bonds with water. This in turn has an adverse effect on the stabilization by solvation. If however, the basicities are measured in solvent where stabilization by solvation is insignificant, then the basicity does increase on going from secondary amine to tertiary amine.

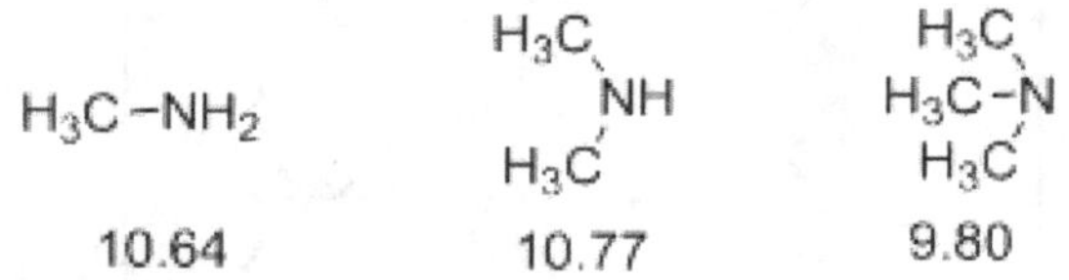

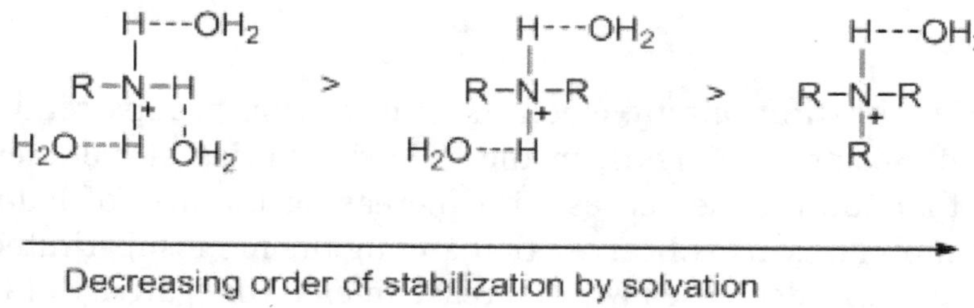

- There is also an effect of stabilization by delocalization in case of strength of bases (like the case of methanoic acid vs ethanoic acid). Thus, it is found that guanidine is much stronger base than amidine which in turn is stronger than urea. The high basic strength is explained by the formation of 3 identical canonical structures of similar energy for guanidium ion. This results in lowering in energy of the molecule and gives stability to the species. A similar case occurs for amidinium ion where there are two such structures of identical energy. In case of urea, there are no such structures of identical energy and thus it is the least basic of the three.

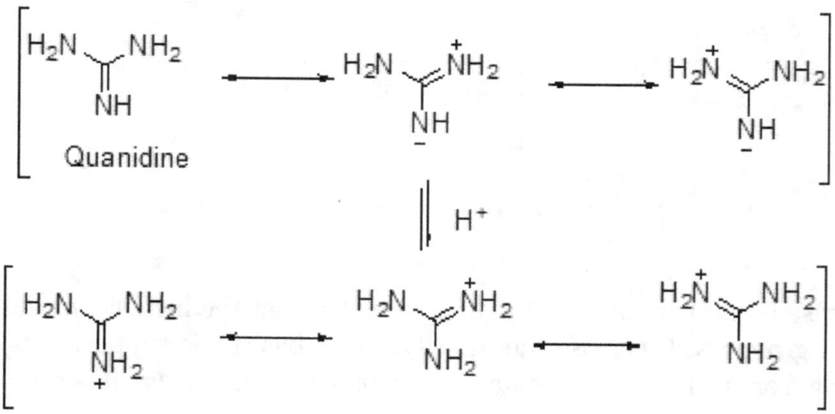

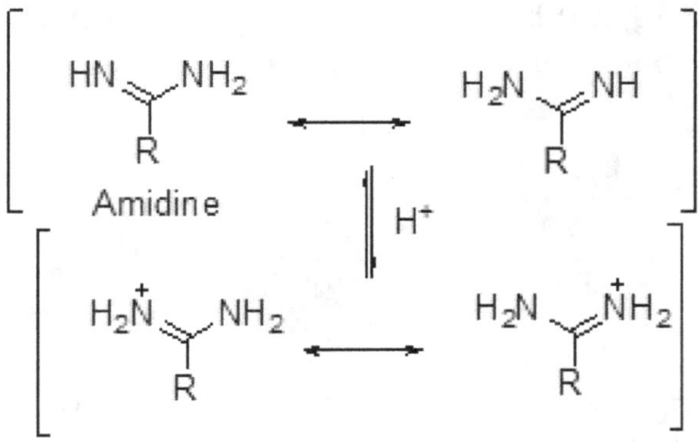

- Electron withdrawing and electron donating mesomeric effects also have influence on the strengths of acids and bases. In the series of *o*- , *m*-, *p*- nitroanilines it is observed that the *ortho* and *para* isomers have extremely high acidity as compared to the *meta* isomer. In fact the *o* -nitroaniline is a stronger acid than *p* -nitroaniline. This is explained in terms of electron withdrawing mesomeric effect which is operational only in the *ortho* and *para* isomers but not in the *meta* isomer. The extra base weakening effect, when the substituent is in *ortho* position in part due to the short distance over which its inductive effect is operating and also to the direct interaction between the nitro and amino group by hydrogen bonding.

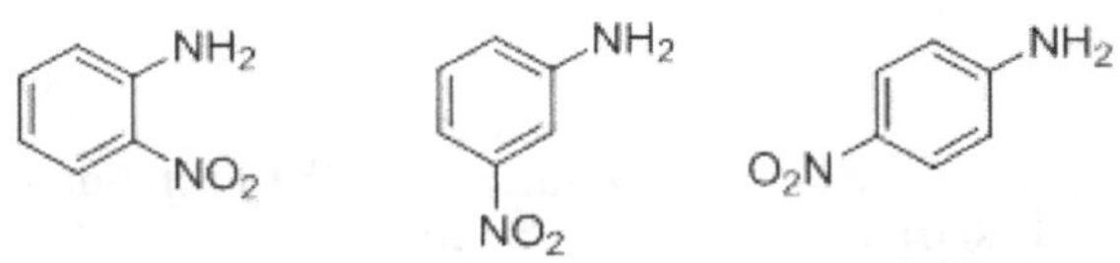

-0.28 2.45 0.98

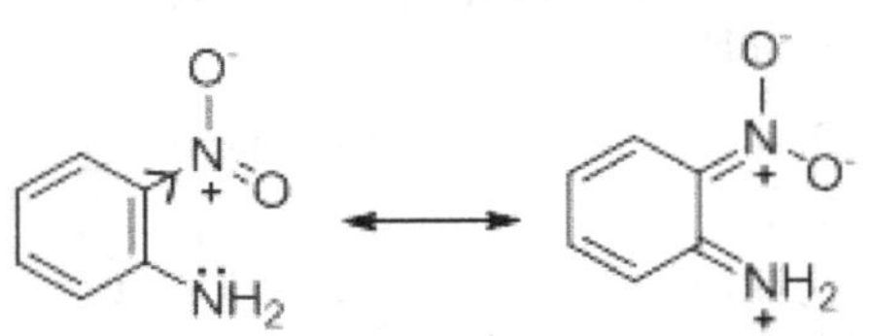

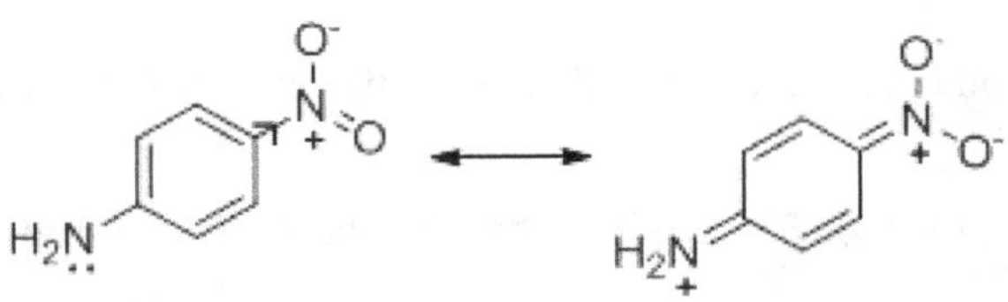

- Steric considerations may also affect the strength of an acid or base. Such an instance is observed in case of *o* -t-butyl benzoic acid which is ten times stronger than its *para* isomer. This is totally unexpected since the electron donating inductive effect should be stronger in the *ortho* isomer compared to the *para*

isomer. However, in this case, the carboxyl group is forced out of the plane of the ring by the bulky t-butyl group in *ortho* position.

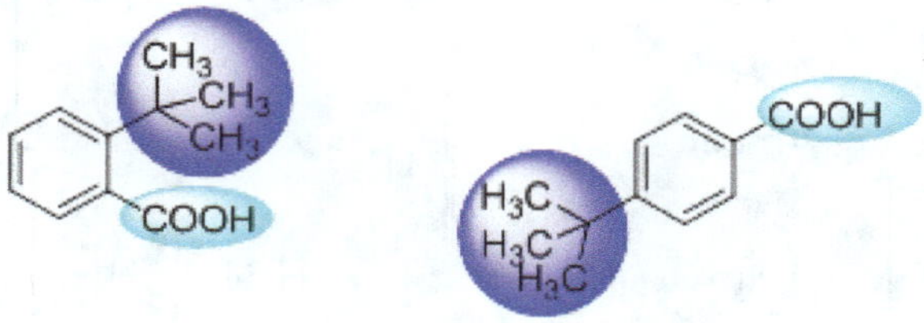

- Steric effects can also be caused by other types of strain. 1,8-Bis(diethylamino)-2,7-dimethoxynaphthalene is an extremely strong base for a tertiary amine, but proton transfers to and from the nitrogen are exceptionally slow; slow enough to be followed by a UV spectrophotometer. This is because the compound is severely strained and the two nitrogen lone pairs are forced to be near each other. Protonation relieves the strain: one lone pair is now connected to hydrogen, which forms a hydrogen bond to the other lone pair.

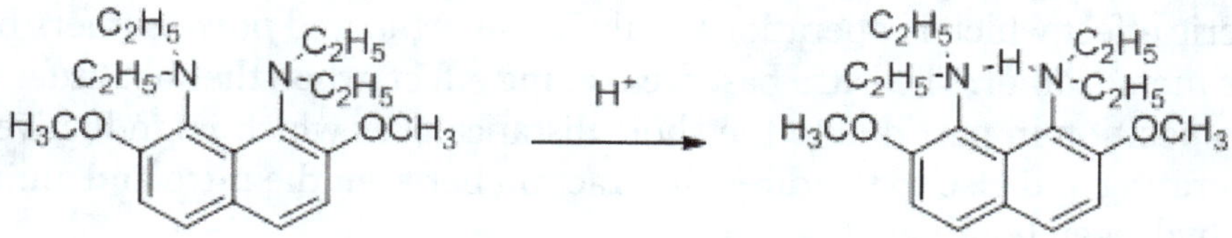

References

- Milne; Scribner; Crawford. "Non-ionic Diffusion and the Excretion of Weak Acids and Bases" (PDF). Science Direct. Retrieved 11 February 2020
- Chemistry, 9th Edition. Kenneth W. Whitten, Larry Peck, Raymond E. Davis, Lisa Lockwood, George G. Stanley. (2009) ISBN 0-495-39163-8. Page 363
- Tanabe, K.; Misono, M.; Ono, Y.; Hattori, H. New Solid Acids and Bases: their catalytic properties. p. 234.
- Ebbing, Darrell; Gammon, Steven D. (2010). General Chemistry, Enhanced Edition. Cengage Learning. pp. 644–645. ISBN 0-538-49752-1
- Ebbing, D.D., & Gammon, S. D. (2005). General chemistry (8th ed.). Boston, MA: Houghton Mifflin. ISBN 0-618-51177-6
- Tanabe, Konzo. Solid Acids and Bases: their catalytic properties. Academic Press. p. 2. Retrieved 12 February 2020
- Jensen, William B. (2006). "The origin of the term "base"" (PDF). The Journal of Chemical Education. 83 (8): 1130
- Reich, Hans J. "Bordwell pKa Table (Acidity in DMSO)". Department of Chemistry, University of Wisconsin, U.S.
- Soil Survey Division Staff. "Soil survey manual.1993. Chapter 3, selected chemical properties.". Soil Conservation Service. U.S. Department of Agriculture Handbook 18

- Pearson, Ralph G. (2005). "Chemical hardness and density functional theory" (PDF). Journal of Chemical Sciences. 117 (5): 369–377. doi:10.1007/BF02708340
- "Oxonium ylides". IUPAC Compendium of Chemical Terminology (interactive version) (2.3.3 ed.). International Union of Pure and Applied Chemistry.
- Masterton, William; Hurley, Cecile; Neth, Edward (2011). Chemistry: Principles and Reactions. Cengage Learning. p. 433. ISBN 1-133-38694-6

Permissions

All chapters in this book are published with permission under the Creative Commons Attribution Share Alike License or equivalent. Every chapter published in this book has been scrutinized by our experts. Their significance has been extensively debated. The topics covered herein carry significant information for a comprehensive understanding. They may even be implemented as practical applications or may be referred to as a beginning point for further studies.

We would like to thank the editorial team for lending their expertise to make the book truly unique. They have played a crucial role in the development of this book. Without their invaluable contributions this book wouldn't have been possible. They have made vital efforts to compile up to date information on the varied aspects of this subject to make this book a valuable addition to the collection of many professionals and students.

This book was conceptualized with the vision of imparting up-to-date and integrated information in this field. To ensure the same, a matchless editorial board was set up. Every individual on the board went through rigorous rounds of assessment to prove their worth. After which they invested a large part of their time researching and compiling the most relevant data for our readers.

The editorial board has been involved in producing this book since its inception. They have spent rigorous hours researching and exploring the diverse topics which have resulted in the successful publishing of this book. They have passed on their knowledge of decades through this book. To expedite this challenging task, the publisher supported the team at every step. A small team of assistant editors was also appointed to further simplify the editing procedure and attain best results for the readers.

Apart from the editorial board, the designing team has also invested a significant amount of their time in understanding the subject and creating the most relevant covers. They scrutinized every image to scout for the most suitable representation of the subject and create an appropriate cover for the book.

The publishing team has been an ardent support to the editorial, designing and production team. Their endless efforts to recruit the best for this project, has resulted in the accomplishment of this book. They are a veteran in the field of academics and their pool of knowledge is as vast as their experience in printing. Their expertise and guidance has proved useful at every step. Their uncompromising quality standards have made this book an exceptional effort. Their encouragement from time to time has been an inspiration for everyone.

The publisher and the editorial board hope that this book will prove to be a valuable piece of knowledge for students, practitioners and scholars across the globe.

Index

www.ingramcontent.com/pod-product-compliance
Lightning Source LLC
LaVergne TN
LVHW080620210926
848485LV00003B/187
* 9 7 8 1 6 3 9 8 9 2 2 9 7 *